The Minerals, Metals & Materials Series

Guillaume Lambotte · Jonghyun Lee
Antoine Allanore · Samuel Wagstaff
Editors

Materials Processing Fundamentals 2018

TMS

Springer

Editors
Guillaume Lambotte
Boston Electrometallurgical Corporation
Woburn, MA
USA

Antoine Allanore
Massachusetts Institute of Technology
Cambridge, MA
USA

Jonghyun Lee
Iowa State University
Ames, IA
USA

Samuel Wagstaff
Novelis
Sierre
Switzerland

ISSN 2367-1181 ISSN 2367-1696 (electronic)
The Minerals, Metals & Materials Series
ISBN 978-3-319-72130-9 ISBN 978-3-319-72131-6 (eBook)
https://doi.org/10.1007/978-3-319-72131-6

Library of Congress Control Number: 2017960289

Printed on acid-free paper

This Springer imprint is published by Springer Nature
The registered company is Springer International Publishing AG
The registered company address is: Gewerbestrasse 11, 6330 Cham, Switzerland

Preface

The symposium Materials Processing Fundamentals is hosted at the annual meeting of The Minerals, Metals & Materials Society (TMS) as the flagship symposium of the Process Technology and Modeling Committee. It is a unique opportunity for interdisciplinary presentations and discussions about, among others, processing, sensing, modeling, multi-physics, computational fluid dynamics, and thermodynamics.

The materials covered include ferrous and nonferrous elements, and the processes range from mining unit operations to joining and surface finishing of materials. Acknowledging that modern processes involve multi-physics, the symposium and its proceedings allow the reader to learn the methods and outcome of other fields modeling practices, often enabling the development of practical solutions to common problems. Modeling of basic thermodynamic and physical properties play a key role, along with computer fluid dynamics and multiphase transport and interface modeling.

Contributions to the proceedings include applications such as steel processing, modeling of steel and nonferrous alloys treatments for properties control, multi-physics, and computational fluid dynamics modeling for molten metal processes and properties measurement. Extractive, recovery, and recycling process modeling are also presented, completing a broad view of the field and practices of modeling in materials processing.

The engagement of TMS and committee members to chair sessions, review manuscript, and help TMS present current practices, makes this symposium and its proceedings possible. The editor and its coeditors acknowledge the invaluable support and contribution of these volunteers as well as TMS staff members, in particular, Patricia Warren, Trudi Dunlap, Carol Matty, and Matt Baker.

Guillaume Lambotte
Jonghyun Lee
Antoine Allanore
Samuel Wagstaff

Contents

About the Editors

Guillaume Lambotte is a Senior Research and Development Scientist at Boston Electrometallurgical Corporation, a Massachusetts Institute of Technology (MIT) spin-off start-up focusing on the development of an environmentally friendly and energetically efficient primary metal extraction process. Dr. Lambotte primarily focuses on computational thermodynamic modeling, electrochemistry, and high-temperature equilibrium. Prior to joining BEMC, he conducted research as a postdoctoral associate at the University of Massachusetts (UMass) Amherst and MIT. Before his graduate studies, Dr. Lambotte worked as a Production Assistant Manager at Alcan Extruded Products (Crailsheim, Germany).

Dr. Lambotte obtained his bachelor degree from the European Engineer School for Materials Science (Nancy, France). He received his M.Sc. and Ph.D. in metallurgical engineering from Ecole Polytechnique of Montreal (Montreal, Canada).

Dr. Lambotte is currently serving as the chair of the TMS Process Technology and Modeling Committee and was the recipient of the 2015 TMS EPD Young Leaders Professional Development Award. In 2015, Dr. Lambotte was one of the TMS representatives at the Emerging Leaders Alliance Conference.

Jonghyun Lee is an Assistant Professor in the Department of Mechanical Engineering at Iowa State University. He has been conducting multiple industry- and government-funded projects in the field of materials processing as PI and Co-I.

Dr. Lee was the recipient of the Young Leaders Professional Development Award in 2013 from The Minerals, Metals & Materials Society, where he has been serving as a co-organizer and coeditor of the Materials Processing Fundamentals Symposium since 2014 and as a vice-chair of the Process Modeling and Technology Committee since 2017.

Prior to joining his current institution, Dr. Lee was a Research Assistant Professor at the University of Massachusetts Amherst. He also had nearly 5 years of industry experience and worked as a postdoctoral associate at Tufts University, Medford, Massachusetts. He earned his M.S. and Ph.D. in mechanical engineering from the University of Massachusetts Amherst and his B.S. in the same discipline from Inha University in Incheon, South Korea.

Antoine Allanore is an Associate Professor of Metallurgy in the Department of Materials Science and Engineering at MIT. He received his higher education in Nancy (France), where he earned a chemical process engineer diploma from Ecole Nationale Supérieure des Industries Chimiques and a M.Sc. and a Ph.D. from Lorraine University.

Dr. Allanore joined MIT in 2012 as a faculty member, leading a research group that develops sustainable materials extraction and manufacturing processes. He has developed numerous alternative approaches for metals and minerals extraction and processing. With an emphasis on electrochemical methods for both analytical and processing purposes, his group combines experimental and modeling approaches to promptly investigate the ultimate state of condensed matter, the molten state. He teaches thermodynamics and sustainable chemical metallurgy at both the undergraduate and graduate level.

He received the Vittorio de Nora Award from TMS in 2012, and the TMS Early Career Faculty Fellow Award in 2015.

Samuel Wagstaff has been working in the aluminum industry since age 14 with Novelis in Spokane, Washington. He received his B.S. from Cornell University in Mechanical and Aerospace Engineering in 2013. He continued his education at the Massachusetts Institute of Technology in the Department of Materials Science and Engineering.

His Ph.D. on the minimization of macrosegregation through jet erosion of a continuously cast ingot uses a turbulent jet to reduce the uneven distribution in aluminum alloy ingots by over 70 %. Dr. Wagstaff finished his masters and doctorate at MIT in September 2016 after just 3 years. He has published more than a dozen articles on DC casting and macrosegregation, and holds 12 patents. He now works for Novelis in Sierre, Switzerland as an Automotive Development and Process Engineer.

Part I
Steelmaking—Processing

The Effect of a Sulfur Addition on the Formation and Behavior of CaS Inclusions During a Secondary Refining Process Without Using a Ca-Treatment

Takanori Yoshioka, Yuta Shimamura, Andrey Karasev, Yasuhide Ohba and Pär Göran Jönsson

Abstract This study aimed to elucidate the effect of a sulfur addition on the formation and behavior of CaS inclusions in steel melts during a secondary refining process without a Ca-treatment. Samples were taken during production for two different steel grades, namely a low-S steel (S = 0.005%) and a high-S steel (S = 0.055%). Thereafter, the inclusion characteristics were determined using an SEM combined with an EDS. The results show that the CaO content in the inclusions decreased and the CaS content increased after a sulfur addition during an RH process for the high-S steel. Furthermore, CaS-covered inclusions were frequently detected in the high-S steel samples after the S addition. Thermodynamic calculations were also performed to compare the CaS formation behavior in the two steels. The results showed that a CaS phase can thermodynamically be formed in the high-S steel melt even without a Ca-treatment. Also, it was indicated that a CaS phase can be formed in two ways, namely a reaction between $\underline{Ca}$ and $\underline{S}$ and a reaction between CaO in inclusions and $\underline{S}$. From the viewpoint of interfacial features, inclusions covered by a CaS phase are thought to possess low contact angles to steel melts. Therefore, CaS-covered inclusions tend to remain in a steel melt. According to the results of this study, CaS inclusions can be formed and deteriorate the castability of high-S containing steels even without a Ca-treatment.

Keywords CaS inclusion · High-S · Ca-treatment · Castability
Thermodynamics · Contact angle

T. Yoshioka (✉) · A. Karasev · P. G. Jönsson
Department of Materials Science and Engineering,
KTH Royal Institute of Technology, Stockholm, Sweden
e-mail: tyoshioka@himeji.sanyo-steel.co.jp

T. Yoshioka · Y. Shimamura · Y. Ohba
Process Research Group, Basic Research Office,
Research & Development Center, Sanyo Special Steel Co., Ltd., Himeji, Japan

G. Lambotte et al. (eds.), *Materials Processing Fundamentals 2018*,
The Minerals, Metals & Materials Series,
https://doi.org/10.1007/978-3-319-72131-6_1

Introduction

Inclusions can be detrimental to a stable casting when they exist in a solid phase at steelmaking temperatures [1]. Generally, solid oxide inclusions have high interfacial energies to steel melts [2]. Therefore, they tend to accumulate on a nozzle wall which can cause a nozzle clogging [1, 3, 4]. In addition to solid oxide inclusions, CaS inclusions are also recognized as being harmful for a high castability since they also exist as a solid phase at steelmaking temperatures [1, 3, 4]. Therefore, the activities of steel components such as $\underline{\mathrm{Ca}}$, $\underline{\mathrm{S}}$, and $\underline{\mathrm{Al}}$ should be controlled carefully when a Ca-treatment is performed [1, 3, 4]. However, a high-S content in a steel product is sometimes required to possess a high machinability. This high-S content leads to a high activity of $\underline{\mathrm{S}}$, which can react with $\underline{\mathrm{Ca}}$ or a CaO phase in a steel melt and generate a CaS phase. However, there are still few discussions on the formation and behavior of CaS inclusions without using a Ca-treatment. From this standpoint, this study aimed to clarify the formation and behavior of CaS inclusions in steel melts without using a Ca-treatment. In practice, steel samples were taken from the ladle during a secondary refining process. Thereafter, the CaS formation and its behavior in a steel melt were discussed.

Experimental Procedures

The procedure of the melt shop is an EAF → LF → RH → CC line. Two steel grades were subject to this study, namely steel A (0.20%C–0.26%Si–0.83%Mn–0.005%S) and steel B (0.36%C–0.76%Si–1.33%Mn–0.055%S) to investigate the effect of the S content on the formation behavior of CaS inclusions in a steel melt. These two steels are manufactured with high basicity slags, which had a composition saturated with both CaO and MgO. During the production of steel B, its high S content was adjusted during an RH treatment since slag/metal reactions are not active during the process. Steel samples were taken at the end of the LF refining (45 min) and the RH treatment (25 min). These samples were named “LF end” and “RH end”, respectively. The compositions of inclusions on the polished cross section of each steel sample were analyzed using an SEM/EDS inclusion analyzer. The scanned area was 100 mm^2. To calculate the content of CaO in the inclusions, the measured small amount of Mn was allotted to an MnS phase and the remaining S was considered to be bound as a CaS phase. After this procedure, the rest of the Ca content was allotted to a CaO phase [5, 6]. The methods for quantitative analyses of steel compositions were the same as presented in the previous work [6]. To discuss variations of inclusion compositions, the $\underline{\mathrm{Ca}}$ content in each steel sample was calculated based on the information of the average inclusion composition and insoluble oxygen content (T.O - $\underline{\mathrm{O}}_{calc}$) at each sampling time. A detailed explanation of this calculation is described in previous papers [7, 8].

Results

Variations of Compositions in the Steel Melts

The variations of the steel compositions during the processes are shown in Fig. 1. The sulfur contents in both steels were reduced below 0.005 mass% at the end of the LF refining process. Thereafter, FeS was added into the steel melts to enable their products to possess designed steel properties. The increase in $\underline{Al}$ in steel B is a result of an Al addition which aimed to compensate $\underline{Al}$ consumption during an RH treatment [6, 9]. The $\underline{Ca}$ contents in the steel melts were around 5 ppm at the end of the LF refining and thereafter decreased below 1 ppm after the following RH treatment. This decrease can be due to vaporization of $\underline{Ca}$ into a gas phase [10], or reactions between $\underline{Ca}$ and other elements during the RH treatment.

Variations of Inclusion Compositions

The variations of the inclusion compositions in each steel are shown in Fig. 2. The compositions of inclusions are plotted in two diagrams: a $CaO–MgO–Al_2O_3$ ternary diagram and a $CaO–CaS–Al_2O_3$ ternary diagram. The open circles in these figures represent the number-averaged composition at each step. At the end of the LF refining, the inclusion compositions were placed on the tie-line connecting the areas of $MgO \cdot Al_2O_3$ and $CaO–Al_2O_{3liq}$ ($CaO–Al_2O_{3liq}$: mass%CaO = 36–58 at 1873 K [11]) in both steels. This result is consistent with the result which has been reported by Yoshioka et al. [6]. The results also showed that some of the inclusions already contained various amounts of CaS at this stage of the LF treatment.

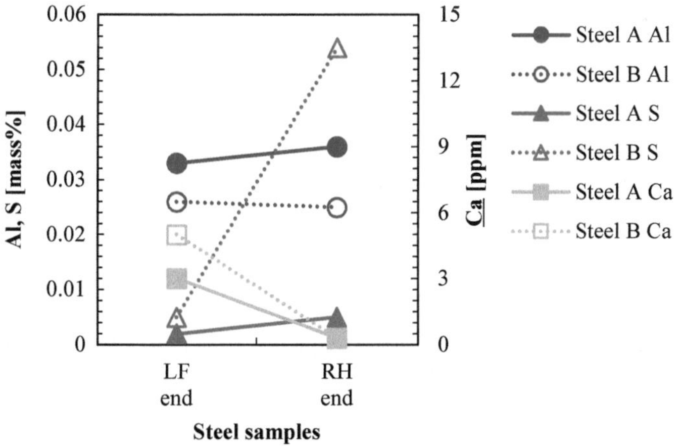

Fig. 1 Variations of the steel compositions in steel A and steel B during the LF-RH process

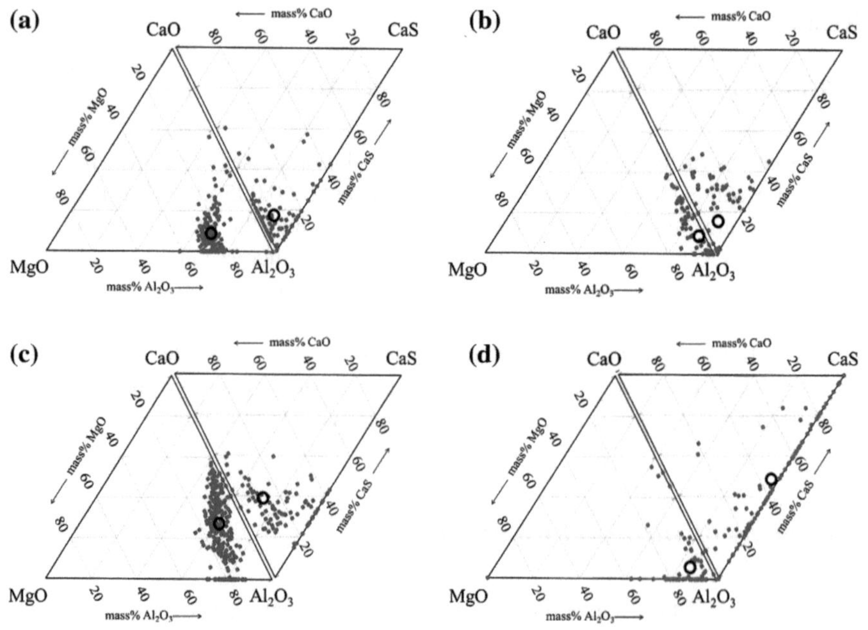

Fig. 2 Inclusion compositions in each sample: **a** LF end of steel A, **b** RH end of steel A, **c** LF end of steel B, **d** RH end of steel B

The difference in the CaS contents in inclusions between the two grades might be due to the difference in the $\underline{Ca}$ and $\underline{S}$ contents during the LF refining process. After the RH treatment, the inclusions in steel A mainly consisted of three phases, namely Al_2O_3, $CaO–Al_2O_{3liq}$, and CaS. On the other hand, the inclusions mainly consisted of two phases, Al_2O_3 and CaS in steel B. In addition to this difference, the CaS contents in the inclusions in steel B were much higher (steel A: 6.9 mass%, steel B: 42.9 mass% in the average composition normalized in the $CaO–MgO–Al_2O_3–CaS$ quarterly system) than those in steel A. Overall, the inclusion compositions were not significantly different between the two steels at the end of the LF refining. However, at the end of the RH treatment, the inclusion compositions were quite different between the two steels with respect to the CaO and CaS contents.

Figure 3 shows the results of SEM observations of the typical inclusion found in each sample. In the samples taken at the end of the LF treatment in both steel grades, the inclusions consisted of two phases, namely a $CaO–Al_2O_3$ phase and an $MgO \cdot Al_2O_3$ phase (Fig. 3a, c). At the end of the RH treatment, Al_2O_3 inclusions were observed in the sample of steel A (Fig. 3b). In the sample of steel B, inclusions surrounded by a CaS phase were frequently observed after the RH treatment (Fig. 3d). These observed results agree with the variations of the inclusion compositions shown in Fig. 2.

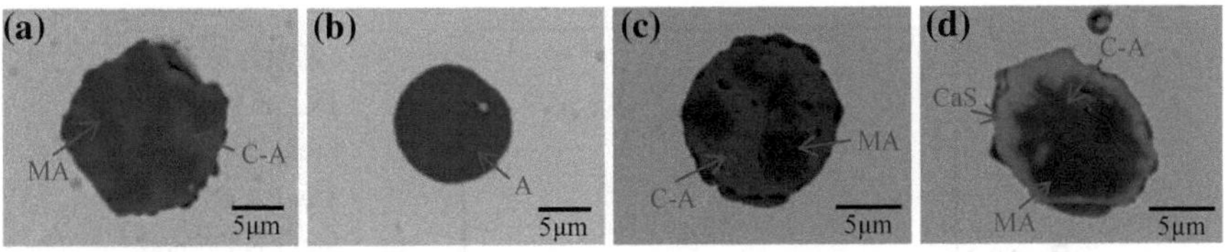

Fig. 3 SEM observations of a typical inclusion found at each stage of the ladle treatment: **a** LF end of steel A, **b** RH end of steel A, **c** LF end of steel B, **d** RH end of steel B

Discussion

Thermodynamic Consideration

As mentioned above, CaS inclusions were observed in the sample of steel B at the end of the RH treatment although no Ca-treatment was used. In the following section, the possible reasons for the CaS formation are discussed from a thermodynamic viewpoint.

CaS Formation by a Reaction Between $\underline{\text{Ca}}$ and $\underline{\text{S}}$ in a Steel Melt

One way to form a CaS phase is due to a reaction between $\underline{\text{Ca}}$ and $\underline{\text{S}}$ in a steel melt. This reaction is expressed in Eq. (1) [12].

$$\underline{\text{Ca}} + \underline{\text{S}} = \text{CaS(s)} \quad \Delta G_1^0 = -530000 + 115.6T\ [\text{J/mol}] \tag{1}$$

A CaS stability diagram was calculated at 1873 and 1823 K, which represent the operation temperatures of the LF refining and the RH treatment, respectively. In this calculation, the interaction parameters shown in Table 1 were used [12–14]. The activities of Ca (a_{Ca}) and S (a_{S}) were calculated with the following Eqs. (2) and (3), where f_i is the activity coefficient of element i in a steel melt.

Table 1 Interaction parameters (e_i^j) of the main elements in the steel melt used in the present study*

j / i	C	Si	Mn	Al	Ca	O
Ca	−0.34	−0.096	−0.0156	−0.072	−0.002	−9000 [14]
Al	0.091	0.056	−0.004	0.043	−0.047	−6.6 [12]
O	−0.45	−0.131	−0.021	−3.9 [12]	−3600 [14]	−0.17
S	0.11	–	−0.026	0.041	−269 [12]	−0.27

*(All data without notation were taken from Ref. [13])

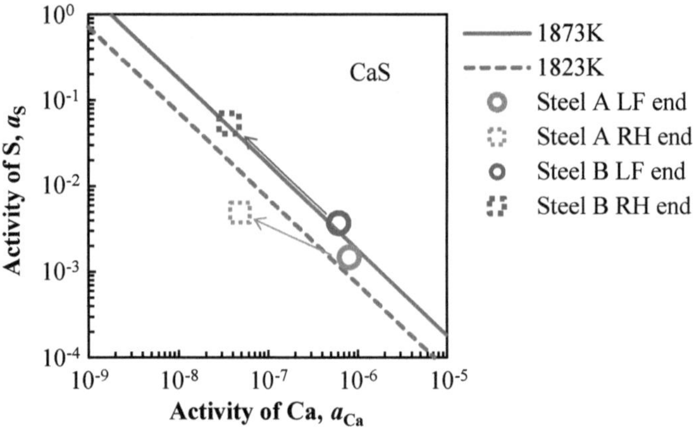

Fig. 4 Stability diagram of CaS formed by a reaction between $\underline{\text{Ca}}$ and $\underline{\text{S}}$ in steel melts

$$a_i = f_i \cdot [\text{mass\% i}] \quad (2)$$

$$logf_i = \sum e_i^j \cdot [\text{mass\% j}] \quad (3)$$

Moreover, the activity of the CaS phase was taken as unity due to its small solubility in CaO–Al_2O_3 phases [15]. Figure 4 shows the obtained relationships between the CaS stability and the activities of $\underline{\text{Ca}}$ and $\underline{\text{S}}$ in steel melts. The symbols in the figure correspond to the Ca and S activities at each sample.

According to this result, $\underline{\text{Ca}}$ and $\underline{\text{S}}$ cannot thermodynamically react to form a CaS phase at the end of the LF refining in any of the two studied steels. Furthermore, CaS cannot be formed in steel A during the RH treatment due to the low Ca and S activities. On the other hand, CaS can be formed in steel B due to the high S activity.

CaS Formation Due to a Reaction Between CaO in Inclusions and $\underline{\text{S}}$ in a Steel Melt

Another way to form a CaS phase in a steel melt is due to a reaction between CaO in inclusions and $\underline{\text{S}}$. This reaction can be expressed as shown in Eq. (5), which was derived by combining Eqs. (1) and (4) [12].

$$\underline{\text{Ca}} + \underline{\text{O}} = \text{CaO(s)} \quad \Delta G_2^0 = -644000 + 148.1T\ [\text{J/mol}] \quad (4)$$

$$\text{CaO(s)} + \underline{\text{S}} + \frac{2}{3}\underline{\text{Al}} = \text{CaS(s)} + \frac{1}{3}\text{Al}_2\text{O}_3 \quad \Delta G_3^0 = -294500 + 98.8T\ [\text{J/mol}] \quad (5)$$

As can be seen in Eq. (5), the progress of this reaction also depends on the activities of CaO and Al_2O_3 in the inclusions. This reaction can easily occur when the CaO activity is high and the Al_2O_3 activity is low. Table 2 shows the activities of CaO and Al_2O_3 at each boundary of the CaO–Al_2O_3 system which was calculated using the thermodynamic software Factsage [6]. These activity data were substituted into Eq. (5) to identify the effect of the inclusion phases. Here, the activity of CaS was set as unity in this calculation.

Figure 5 shows the calculated relationships between the activities of $\underline{Al}$ and $\underline{S}$ to form a CaS phase by the reaction of Eq. (5). The plots in the figure correspond to the Al and S activities at each sample. As has been frequently reported, a CaO–Al_2O_{3liq} phase is the most stable oxide in an Al-killed steel melt [6, 16–18]. Based on the result shown in Fig. 5, the CaO–Al_2O_{3liq} phase can coexist with a CaS phase under a low S activity condition. In the current study, this condition prevails at the end of the LF treatment for both steels and at the end of the RH treatment for steel A. At the end of the RH treatment in steel B, the activity of $\underline{S}$ is greatly increased from 0.004 to 0.05 by the FeS addition. Therefore, the modified CaO–Al_2O_3 phases, such as CaO · Al_2O_3 and CaO–Al_2O_{3liq}, can react with $\underline{S}$ to form a CaS phase. This reaction contributes to both a decrease in the CaO contents and an increase in the CaS contents in inclusions in a steel melt, which were seen in Figs. 2c, d.

Table 2 Activities of CaO and Al_2O_3 in various boundaries of a CaO–Al_2O_3 system at 1873 K [6]

Boundary	a_{Al2O3}	a_{CaO}
Al_2O_3(s)/CaO·6Al_2O_3(s)	1.0	0.0049
CaO·6Al_2O_3(s)/CaO·2Al_2O_3(s)	0.88	0.010
CaO·2Al_2O_3(s)/CaO·Al_2O_3(s)	0.29	0.10
CaO·Al_2O_3(s)/CaO–Al_2O_3(l)	0.18	0.17
CaO–Al_2O_3(l)/CaO(s)	0.0089	0.99

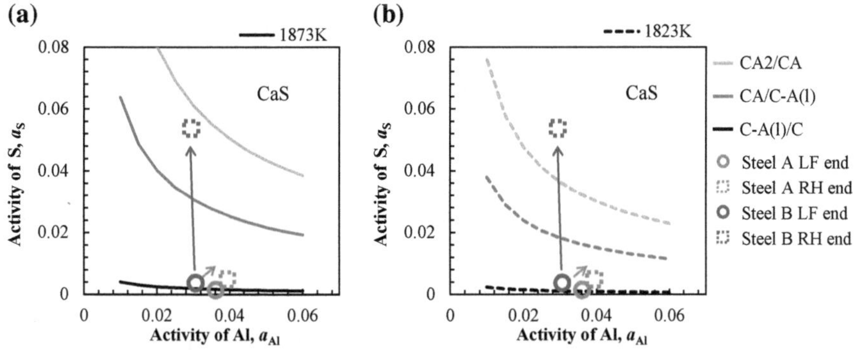

Fig. 5 Relationships between the activities of Al and S to form a CaS phase by a reaction between CaO in various CaO–Al_2O_3 phases and $\underline{S}$ in steel melts at **a** 1873 K and **b** 1823 K. (C: CaO, A: Al_2O_3)

Overall, the results in this study indicate that a CaS phase can be formed in a steel melt even when a Ca-treatment has not been used. The reactions for this CaS formation can progress in two manners, namely due to a reaction between $\underline{Ca}$ and $\underline{S}$ and due to a reaction between CaO in inclusions and $\underline{S}$. These reactions are thought to progress at the inclusion/metal interface. This is the explanation to why the inclusions in the production of steel B were covered by a CaS phase after the FeS addition.

Behavior of CaS Inclusions in a Steel Melt

The removal tendency of an inclusion from a steel melt is greatly affected by an interfacial property, namely the contact angle between an inclusion and a steel melt [19–21]. Arai et al. have clearly elucidated the effect of contact angles on the removal behavior of particles in a liquid [19]. According to their work, the removal tendency steeply decreases when the contact angle between an inclusion and a steel becomes lower than 90°. Generally, solid oxides such as Al_2O_3, CaO–Al_2O_3 with low CaO contents, and MgO · Al_2O_3 have large contact angles in contact to steel melts (>90°) [2]. This information means that solid oxides are easy to remove from a melt. On the other hand, liquid oxides, such as CaO–Al_2O_{3liq}, have small contact angles (50°–60° [2]), so they are difficult to remove from a steel melt. The contact angle of a CaS phase to a steel melt has been reported as 87° [22]. The $\underline{S}$ content, mentioned in ref. [22], in the steel melt is 0.01 mass%. Sulfur is well-known to be a surface active element, which can decrease the interfacial energies of slag/metal and inclusion/metal interfaces [23, 24]. By considering the sulfur content of high-S steels (around 0.05 mass% as shown in Fig. 1), the contact angles of a CaS phase to the steel melts are thought to be 87° or less. Therefore, it can be implied that CaS-covered inclusions, which were identified in Figs. 2d and 3d, are difficult to remove, and they tend to remain in a steel melt even after the completion of an RH treatment.

As mentioned above, liquid inclusions and CaS inclusions are difficult to remove from a steel melt. However, liquid inclusions rarely cause a nozzle clogging even if they remain in a steel melt after refining processes. On the other hand, CaS inclusions are well-known to deteriorate the castability since they exist as solid inclusions in a steel melt (melting point $\approx$ 2800 K) [1, 3, 4, 11]. Thus, CaS inclusions have quite undesirable characteristics such as a tendency to remain in a steel melt and a tendency to accumulate on a nozzle wall, which can causes a nozzle clogging. Table 3 summarizes the results of this discussion.

A deposition on a nozzle wall of a ladle after casting steel B was investigated using an SEM in combination with an EDS. The result is shown in Fig. 6. As seen, sulfides were frequently detected in the deposition. Thus, CaS inclusions can cause a clogging in production of high-S steels even for cases where a Ca-treatment has not been used.

Table 3 Dominant inclusion types and their behavior in low-S steel melts and in high-S steel melts

Inclusion type	Removal from a melt	Clogging	In low-S steels without a Ca-treatment	In high-S steels without a Ca-treatment
Solid oxide*	Preferable (easy)	Unpreferable (can cause)	Dominant (Clogging cannot occur if solid inclusions are properly removed)	–
Liquid oxide*	Unpreferable (difficult)	Preferable (cannot cause)		
CaS inclusion	Unpreferable (difficult)	Unpreferable (can cause)	–	dominant (CaS inclusions tend to remain in steel melts and cause cloggings)

*At a steelmaking temperature

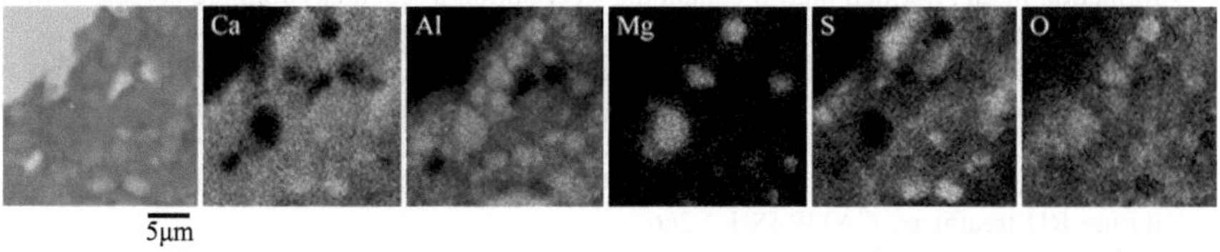

Fig. 6 Element mappings of a deposition on a ladle nozzle in the production of steel B

Conclusion

Plant experiments were carried out to study the formation and behavior of CaS inclusions in a steel melt during a secondary refining process without using a Ca-treatment. The inclusion characteristics in the steel samples taken at the end of an LF treatment and an RH treatment were determined using an SEM in combination with an EDS. Furthermore, thermodynamic calculations were performed to discuss the possibilities of forming CaS inclusions in steels without a Ca-treatment. Also, the influence of the interfacial properties between the inclusions and steel melts on the behavior of the observed inclusions in the steel melts was discussed. Based on the results of this study, the following conclusions can be drawn.

In a high-S steel melt, a CaS phase can be formed even when a Ca-treatment has not been used. Specifically, this can take place in two manners: a reaction between $\underline{Ca}$ and $\underline{S}$, and a reaction between CaO in inclusions and $\underline{S}$.

Due to the formation of a CaS phase, inclusions in high-S containing steel melts are covered by a CaS layer, which is difficult to remove from the melts. Therefore, the castability of high-S containing steels can be deteriorated by a deposition of CaS inclusions even without a Ca-treatment.

References

1. Holappa L (2001) Inclusion control for castability of resulphurized steels. In: Proceedings of the annual steelmaking conference, Warrendale, USA, pp 765–777
2. Cramb AW, Jimbo I (1989) Calculation of the interfacial properties of liquid steel—slag systems. Steel Res Int 60:157–165
3. Janke D, Ma Z, Valentin P, Heinen A (2000) Improvement of castability and quality of continuously cast steel. ISIJ Int 40:31–39
4. Fruehan RJ (1997) Future steelmaking technologies and the role of basic research. Metall Mater Trans A 28:1963–1973
5. Kawakami K, Taniguchi T, Nakashima K (2007) Generation mechanisms of non-metallic inclusions in high-cleanliness steel. Tetsu-to-Hagané 93:743–752
6. Yoshioka T, Nakahata K, Kawamura T, Ohba Y (2016) Factors to determine inclusion compositions in molten steel during the secondary refining process of case-hardening steel. ISIJ Int 56:1973–1981
7. Yoshioka T, Shimamura Y, Karasev A, Ohba Y, Jönsson P (2017) Mechanism of a CaS formation in an Al-killed high-S containing steel during a secondary refining process without a Ca-treatment. Steel Res Int 87, online published (srin. 201700147)
8. Yoshioka T, Ideguchi T, Karasev A, Ohba Y, Jönsson P (2017) The effect of a high Al content on the variation of the total oxygen content in the steel melt during a secondary refining process. Steel Res Int 87, online published (srin. 201700287)
9. Higuchi Y, Shirota Y, Obana T, Ikenaga H (1991) Behavior of inclusion in molten steel during RH treatment. CAMP-ISIJ 4:266
10. Mikhailov GG, Baibulenko EP (1981) Thermodynamics of interaction processes of oxygen and sulphur with calcium in liquid iron. Steel USSR 11:443–444
11. Verein Deutscher Eisenhuttenleute (1995) Slag Atlas, 2nd edn. Verlag Stahleisen GmbH Dusseldorf, Germany
12. Inoue R, Suito H (1994) Calcium desulfurization equilibrium in liquid iron. Steel Res Int 65:403–409
13. Hino M, Ito K (2010) Thermodynamic data for steelmaking. Tohoku University Press, Sendai
14. Cho SW, Suito H (1994) Assessment of calcium-oxygen equilibrium in liquid iron. ISIJ Int 34:265–269
15. Fujisawa T, Inoue S, Takagi S, Wanibe Y, Sakao H (1985) Solubility of CaS in the molten CaO-Al_2O_3-CaS slags and the equilibrium between the slags and molten steel. Tetsu-to-Hagané 71:839–845
16. Todoroki H, Mizuno K (2003) Variation of inclusion composition in 304 stainless steel deoxidized with aluminum. ISS Trans I&SM 30:60–67
17. Deng Z, Zhu M (2013) Evolution mechanism of non-metallic inclusions in Al-killed alloyed steel during secondary refining process. ISIJ Int 53:450–458
18. Yang W, Zhang L, Wang X, Ren Y, Liu X, Shan Q (2013) Characteristics of inclusions in low carbon Al-killed steel during ladle furnace refining and calcium treatment. ISIJ Int 53:1401
19. Arai H, Matsumoto K, Shimasaki S, Taniguchi S (2009) Model experiment on inclusion removal by bubble flotation accompanied by particle coagulation in turbulent flow. ISIJ Int 49:965–974
20. Zheng X, Hayes PC, Lee H-G (1997) Particle removal from liquid phase using fine gas bubbles. ISIJ Int 37:1091–1097
21. Cho J-S, Lee H-G (2001) Cold model study on inclusion removal from liquid steel using fine gas bubbles. ISIJ Int 41:151–157

22. Staronka A, Gotas W (1979) Investigations into the wettability of solid oxide and sulphide phases with pure iron melts and stainless steel melts. Arch Eisenhüttenw 50:237–242
23. Ogino K (1975) Interfacial tension between molten iron alloys and molten slags. Tetsu-to-Hagané 61:2118–2132
24. Poirier DR, Yin H, Suzuki M, Emi T (1998) Interfacial properties of dilute Fe-O-S melts on alumina substrates. ISIJ Int 38:229–238

Desulfurization of Copper-Iron Reduced from Copper Slag

Bao-jing Zhang, Ting-an Zhang, Li-ping Niu, Zhi-he Dou, Zhi-qiang Li and Dong-liang Zhang

Abstract In order to maximize the use of copper slag, a new idea that copper slag is reduced to smelt copper-containing antimicrobial stainless steel was proposed. But copper-iron reduced from copper slag contains a large number of copper matte, making sulfur content high. In this article, desulfurization of copper-iron was studied. The Fact-Sage software was used to calculate ΔG of the desulfurization reaction. Calcium oxide, calcium carbide and ferromanganese were used as desulfurization agent. The results show that desulfurization capacity of calcium oxide is poor, but with addition of carbon, desulfurization effect of calcium oxide will be enhanced. Calcium carbide and ferromanganese have good desulfurization effect.

Keywords Desulfurization · Copper-iron · Calcium oxide · Calcium carbide Ferromanganese

Introduction

With the development of copper smelting process, strong oxidation process is put into practice. Copper matte grade is higher, and copper element in smelting slag is also higher [1]. At present, the most common use of copper slag is that copper slag is diluted to obtain copper matte and tailings. Matte is back to copper smelting process, and tailings are made into cement, which make copper slag used with low value [2–7]. Based on this, our group proposed a new technology that copper slag is reduced to smelt copper containing antimicrobial stainless steel. That is, copper slag is directly reduced to obtain copper-containing molten iron, and then by a series of

---

B. Zhang · T. Zhang (✉) · L. Niu · Z. Dou · Z. Li · D. Zhang
Key Laboratory of Ecological Metallurgy of Multi-metal Intergrown Ores of Ministry of Education, Special Metallurgy and Process Engineering Institute, Northeastern University, Shenyang 110819, China
e-mail: zta2000@163.net

G. Lambotte et al. (eds.), *Materials Processing Fundamentals 2018*,
The Minerals, Metals & Materials Series,
https://doi.org/10.1007/978-3-319-72131-6_2

treatment desulfurization, decarburization and alloying to get copper-containing antimicrobial stainless steel [8].

Copper element in smelting slag is mostly in form of matte, and part of copper element is still in form of matte in reduction process [9]. High sulfur content in copper-containing molten iron affects the mechanical properties of the final product. The desulfurization of copper-containing molten iron is a key issue. In steelmaking process, the desulfurizers are mainly calcium oxide, calcium carbide and magnesium, which are used to remove dissolved sulfur in molten iron [10, 11]. Sulfur content is high in slag and most exists in form of cuprous sulfide. The desulfurization effect of traditional desulfurization agent may be different. In this article, desulfurization effect of calcium oxide, calcium carbide and ferromanganese were studied.

Experiment

The raw materials used in the experiment are copper and iron alloys obtained from copper slag. The main compositions of alloy (wt%) are as follows: Cu, 9.05%; Fe, 84.92%; S, 1.70%. The main components of alloy are copper and iron, and copper matte is mixed in iron matrix. The experiments were carried out in an induction furnace. Alloy is placed in an alumina crucible, and an external graphite crucible is subjected to auxiliary heat. The experiment device is shown in Fig. 1. Calcium silicate slag (5 g calcium oxide, 5 g SiO_2) is used as protection slag to avoid oxidation of metal. Three experiments are done to study the desulfurization effect. Calcium oxide, calcium carbide and ferromanganese are as desulfurization agent in metal separately. The reaction time is 1450 °C and reaction time is 30 min.

The reactions thermodynamics are calculated by FactSage software,database (FactPS). The composition of metal after desulfurization are analyzed. Cu, Fe and Mn in metal were detected by the atomic absorption spectrometer. S was detected

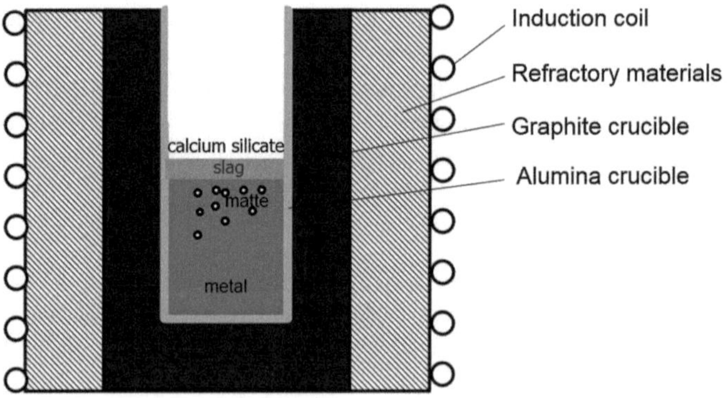

Fig. 1 Experimental setup

by the carbon and sulfur analyzer. The morphology metal samples were detected with scanning electron microscope SU8000 (voltage 20 kV, 20 μA).

Results and Analysis

Calcium Oxide

Metal raw material is 100 g. Calcium oxide (5 g) and silica (5 g) are as protection slag, and calcium oxide (5 g) is added as desulfurizer. After experiment, quality of metal obtained is 99.51 g and metal components (wt%) are as follows: Cu, 8.77%, Fe, 84.98%, S, 1.68%. Sulfur removal rate is very low.

The desulfurization principle of CaO is that solid CaO absorbs sulfur in molten iron. Silicon and carbon in molten iron are excellent reducing agents, which can absorb oxygen produced by reaction. Desulfurization reaction of CaO with dissolved sulfur in molten iron is as follows [12]:

$$CaO + [S] + [C] = CaS + CO(g) \tag{1}$$

When [Si] > 0.05%, the following reaction occurs

$$CaO + [S] + 0.5[Si] = CaS + 0.5Ca_2SiO_4 \tag{2}$$

The product of desulfurization reaction is CaS and CO (or SiO_2). High temperature, high alkalinity, reducing atmosphere and [C], [Si], [P] with high sulfur activity coefficient in molten iron are beneficial to desulfurization.

Sulfur in copper-containing molten iron obtained from copper slag is present in form of copper matte, that is, cuprous sulfide, ferrous sulfide and a combination of both. The reactions of calcium oxide and sulfide is as follows:

$$CaO + FeS = CaS + FeO \tag{3}$$

$$CaO + FeS + 1/2C = CaS + Fe + 1/2CO_2(g) \tag{4}$$

$$CaO + FeS + C = CaS + Fe + CO(g) \tag{5}$$

$$CaO + Cu_2S + C = CaS + Cu + CO(g) \tag{6}$$

$$CaO + Cu_2S + 1/2C = CaS + Cu + 1/2CO_2(g) \tag{7}$$

$$CaO + Cu_2S = CaS + Cu_2O \tag{8}$$

The thermodynamic calculations (1473–2073 K) of the reaction of calcium oxide and sulfide are performed. Components have solids {CaO, C, CaS, Fe (<1810.95 K), Cu_2O(<1516.70 K), FeO(<1644.15 K)}, liquids{FeS, Cu_2S, Fe

(>1870.95 K), Cu, Cu_2O(>1516.70 K), FeO(>1644.15 K)} and gases (CO_2, CO). The results are shown in Fig. 2.

It can be seen from Fig. 2 that in the presence of calcium oxide alone, only ferrous sulfide can react with calcium oxide. After carbon is added, cuprous sulfide is reduced and desulfurization reaction of ferrous sulfide is easier. Carbon addition makes desulfurization capacity of calcium oxide greatly increased. Studies have shown that the main limiting links of CaO desulfurization are the contact area of calcium oxide with molten iron and the product layer on CaO surface [13]. If silica content is too high, calcium oxide will react with it to produce $2CaO{\cdot}SiO_2$, which is dense layer on the surface of calcium oxide, affecting subsequent reaction of desulfurization. In this experiment, calcium silicate slag is as mold powder, which is not conducive to the desulfurization of reaction. Alumina crucible will react with calcium oxide, increasing the difficulty of desulfurization. In the conditions without carbon, sulfur removal rate is low. The scanning electron microscopy analysis of metal product in experiment is shown in Fig. 3. A large number of matte is in copper and iron matrix. Although CaO desulfurizer has some shortcomings, it has

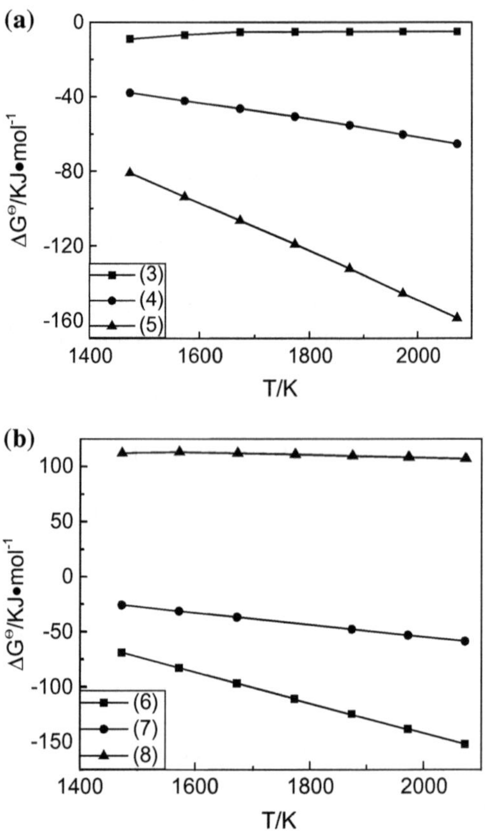

Fig. 2 $\Delta G^{\ominus}$ variation with temperature: **a** CaO with FeS; **b** CaO with Cu_2S

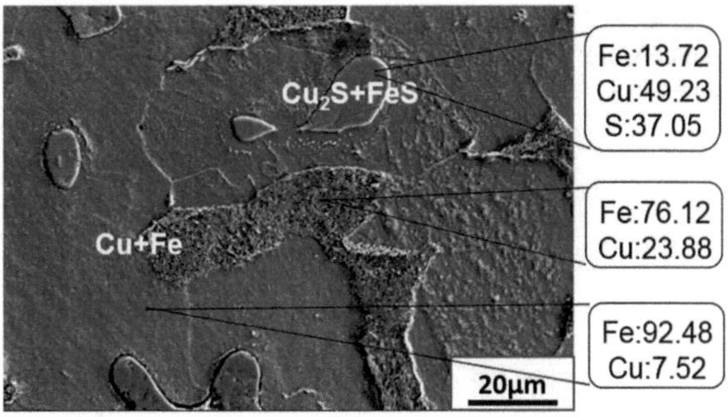

Fig. 3 Electron microscopy results of metal in experiment 1

been widely used, especially by the emergence of the Kanbara reactor method. Reduction conditions should be ensured in desulfurization.

Calcium Carbide

Metal raw material is 100 g. Calcium oxide (5 g) and silica (5 g) are as protection slag, and calcium carbide (8 g) is added as the desulfurizer. After experiment, quality of metal obtained is 96.35 g and metal components (wt%) are as follows: Cu, 9.25%, Fe, 87.65%, S, 0.028%. Sulfur removal rate is 98.41%.

CaC_2 desulfurization is solid-liquid reaction, and desulfurization reaction of CaC_2 with sulfur dissolved in molten iron is as follows:

$$[S] + CaC_2 = CaS + 2C(s) \tag{9}$$

When the reaction is balanced, the equilibrium sulfur content in molten iron is much lower than in actual production. CaC_2 is a desulfurizer with the strongest desulfurization ability. Studies have shown that desulfurization limits are the size of CaC_2 particles and the diffusion rate of sulfur in molten metal side boundary layer. Usually, the desulfurization reaction rate can be improved by the methods that CaC_2 particle size is reduced, reaction area is increased and the stirring is enhanced.

The reactions of calcium carbide and sulfide are as follows:

$$CaC_2 + Cu_2S = CaS + 2Cu + 2C \tag{10}$$

$$CaC_2 + FeS = CaS + Fe + 2C \tag{11}$$

The thermodynamic calculations of the reaction of calcium carbide and sulfide are performed. Components have solids {CaC_2, C, CaS, Fe(<1810.95 K)} and liquids {FeS, Cu_2S, Fe(>1870.95 K), Cu}. The results are shown in Fig. 4. It can be seen from Fig. 4 that CaC_2 can react with ferrous sulfide and cuprous sulfide. The scanning electron microscopy analysis of metal product in experiment is shown

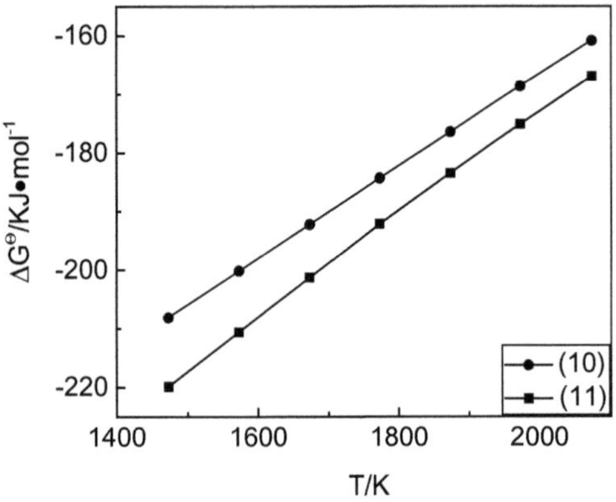

Fig. 4 ΔG^{Θ} variation with temperature: CaC_2 with FeS and Cu_2S

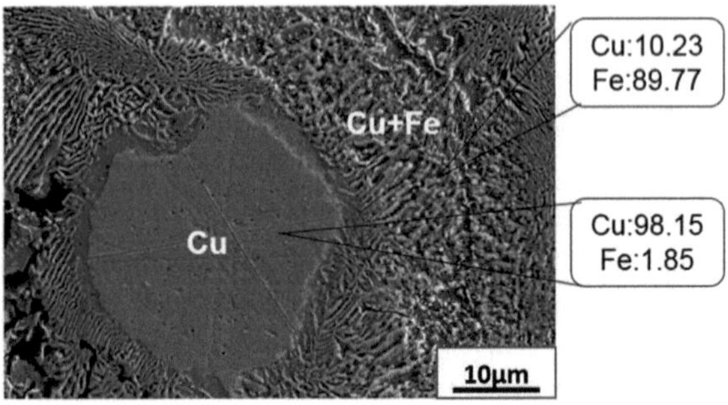

Fig. 5 Electron microscopy results of metal in experiment 2

in Fig. 5. It can be known that only copper mixed in the copper and iron matrix, and the place copper exists is the place original matte exists, indicating matte is reduced.

Calcium carbide is not easy to save, easy to hydrolyze, and difficult to transport. In actual production, there are security problems when using calcium carbide desulfurization. Nowadays, calcium carbide is generally used as a composite desulfurization agent. But its strong desulfurization can not be ignored.

Ferromanganese Alloy (Fe-65 wt%Mn)

Metal raw material is 100 g. Calcium oxide (5 g) and silica (5 g) are as protection slag, and ferromanganese alloy (10.56 g) is added as the desulfurizer. After experiment, quality of metal obtained is 106.69 g and metal components (wt%) are as follows: Cu, 8.46%, Fe, 83.40%, S, 0.258%, Mn, 2.43%. Sulfur removal rate is 83.81%.

Desulfurization reaction of ferromanganese alloy with sulfur dissolved in molten iron is as follows:

$$Mn + [S] = MnS \quad (12)$$

The desulfurization component of ferromanganese is manganese. Manganese desulfurization is mainly using the property that binding capacity of metal manganese and sulfur is far greater than iron and copper. The reactions of ferromanganese alloy and sulfide are as follows:

$$Mn + FeS = MnS + Fe \quad (13)$$

$$Mn + Cu_2S = MnS + 2Cu \quad (14)$$

The thermodynamic calculations of the reaction of manganese and sulfide are performed. Components have solids {Mn(<1519 K), MnS(<1928.05 K), Fe (<1810.95 K)} and liquids{FeS, Cu_2S, Fe(>1870.95 K), Cu, Mn(>1519 K), MnS (>1928.05 K)}. The results are shown in Fig. 6.

It can be seen from Fig. 6 that manganese can react with ferrous sulfide and cuprous sulfide. Desulfurization of ferromanganese alloy is efficient, deep, producing slag less and making temperature dropping less. Manganese has a certain solubility in molten iron, which can prevent the return of sulfur. The scanning electron microscopy analysis of metal product in experiment is shown in Fig. 7. It can be known that metal consists of three parts, copper - iron - manganese alloy matrix, copper alone and manganese sulfide inclusions. Copper sulfide and ferrous sulfide are all reduced by manganese, but the resulting manganese sulfide is present in metal. It can be seen from Fig. 7 that copper matte is gradually replaced by manganese, indicating manganese has a strong desulfurization effect. But manganese sulfide inclusions should be avoided. Manganese is usually added into molten iron with the form of ferromanganese. Manganese has a strong adaptability to raw materials, but it also will introduce manganese into metal. It is suitable for smelting manganese steel.

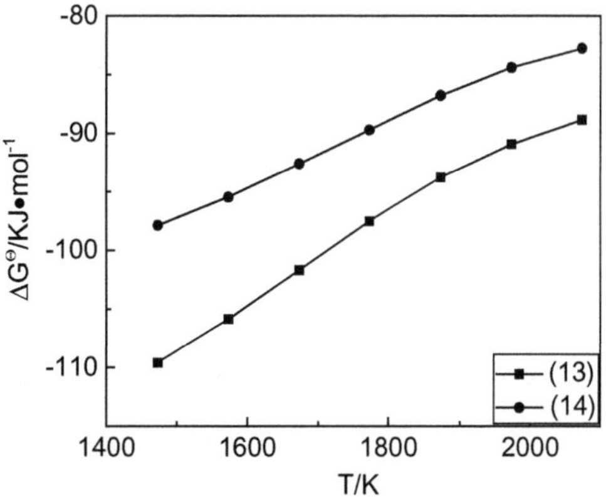

Fig. 6 $\Delta G^{\ominus}$ variation with temperature: Mn with FeS and Cu_2S

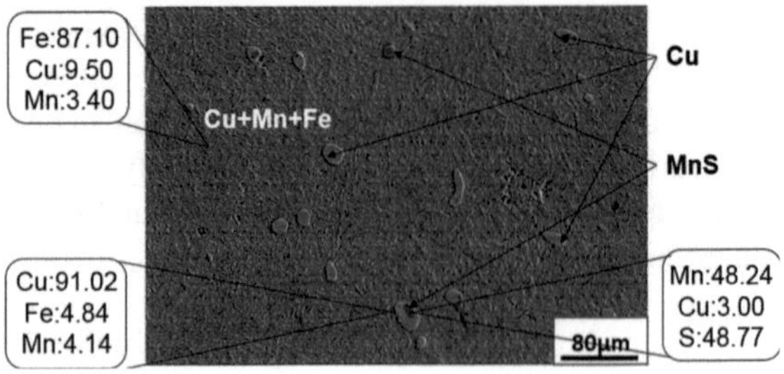

Fig. 7 Electron microscopy results of metal in experiment 3

Conclusions

A large number of copper matte is in metal obtained from copper slag. Calcium oxide, calcium carbide and ferromanganese were used to research desulfurization effect, and desulfurization mechanism was analyzed. conclusion are as follows:

1. Calcium oxide and copper matte is difficult to react and calcium oxide alone has poor desulfurization effect. The metal after desulfurization is mixed with a large number of unreacted cuprous sulfide and ferrous sulfide. Desulfurization rate of experiment 1 is very low. When carbon added, sulfur removal rate will be improved.
2. Calcium carbide can react with ferrous sulfide and copper sulfide, and desulfurization effect is good. The main component of the metal after desulfurization is copper-iron alloy. The area only containing copper element is formed by desulfurization of matte. Desulfurization rate of experiment 2 is 98.41%. Calcium sulfide is suitable for the removal of sulfur in alloy obtained by copper slag.
3. Manganese can react with ferrous sulfide and copper sulfide. After desulfurization, copper sulfide and ferrous sulfide phase disappear, indicating that desulfurization effect is good. But manganese sulfide is produced and metal is not easy to separate with it. Metal is also mixed with manganese. Desulfurization rate of experiment 3 is 83.81%. During production process, the amount of ferromanganese added should be control, and the physical properties of slag should be improved, making manganese sulfide easier into slag.

Acknowledgements This research was supported by the Basic Research Universities Special Fund Operations (N140204013, N130102002, N130702001); the Education Department of Liaoning Province (LZ2014021); National Natural Science Foundation of China (51374064).

References

1. Cao H, Zhang L, Fu N et al (2009) Review of copper slag impoverishment. J Mater Metall 8:33–39
2. Gorai B, Jana RK, Premchand (2003) Characteristics and utilisation of copper slag—a review. Resour Conserv Recycl 39:299–313

3. Murari K, Siddique R, Jain KK (2015) Use of waste copper slag, a sustainable material. J Mater Cycles Waste Manage 17:13–26
4. Mostafa K, Ali B (2009) Mechanical properties of high-strength concrete incorporating copper slag as coarse aggregate. Constr Build Mater 23:2183–2188
5. Wei W, Zhang W, Ma G (2010) Optimum content of copper slag as a fine aggregate in high strength concrete. Mater Des 31:2878–2883
6. Shi C, Meyer A Behnood (2008) Utilization of copper slag in cement and concrete. Resour Conser Recycl 2008(52):1115–1120
7. Moskalyk RR, Alfantazi AM (2003) Review of copper pyrometallurgical practice: today and tomorrow. Miner Eng 2003(16):893–919
8. Zhang T, Niu L, Dou Z et al. (2014) A method molten iron reduced from copper slag directly smelt copper antimicrobial stainless steel. Chinese Patent, Patent No. CN 1041203510 A, 2014
9. Zhang B, Niu L, Zhang T et al (2017) Alternative reduction of copper matte in reduction process of copper slag. ISIJ Int 57:775–781
10. Lindstrom D, Sichen D (2015) Study on desulfurization abilities of some commonly used desulfurization agents. Steel Res Int 86:73–83
11. Hao X, Wang X, Wang W (2015) Study on desulfurization ability of $CaO-Al_2O_3-SiO_2$ and $CaO-CaF_2$ slags at 1600 degrees C. Steel Res Int 86:1455–1460
12. Wu Y (1996) elementary introduction of the thermodynamics and kinetics in the process of desulphurization. Steelmaking 2:44–52
13. Feng J, Zhao W, Ma G et al. (2009) Analysis of desulphurization of injection and stirring for hot metal pretreatment. Metall Equip, 32–35

Part II
Steelmaking—Properties

Effects of Aging Treatment on the Microstructure and Mechanical Properties of a Nanoprecipitates-Strengthened Ferritic Steel

Y. Zhao, Y. Cui, H. Guo, S. S. Xu, X. H. Wei and Z. W. Zhang

Abstract Aging treatment is the primary method to control the strength of nanoprecipitates-strengthened ferritic steels and excellent mechanical properties could be obtained by optimal combination of the size and the density of nanoclusters. In this study, the aging treatments with different time and temperatures are used to control the microstructure, size and density of nanophases. In addition, the distribution of the nanophases and dislocation pattern are observed. It is found that the strength shows an obvious transformation with the aging time and temperatures. However, the microstructures and ductility have not significantly changed with altering aging process. Furthermore, the relationship between the nature of nanophase and mechanical properties is discussed.

Keywords Microstructure · Mechanical properties · Aging treatment
Ferritic steel

Introduction

Nano-precipitates-strengthened ferritic steels are becoming increasingly important in bridging industry, power generation, shipbuilding and automotive industries [1–3]. For traditional carbon steel, the increasing of strength is mainly dependent on martensite, which significantly reduces the ductility and toughness of steels [4, 5]. However, the main strengthening mechanisms of nanophase strengthened ferritic steels are precipitation strengthening [6]. The different degree of strengthening is highly dependent on the type, number density, size and spatial distribution of the precipitates [7]. The nature of nanoprecipitates can be controlled by aging

Y. Zhao · Y. Cui · H. Guo · S. S. Xu · X. H. Wei · Z. W. Zhang (✉)
Key Laboratory of Superlight Materials and Surface Technology, Ministry of Education,
College of Materials Science and Chemical Engineering, Harbin Engineering University,
Harbin 150001, People's Republic of China
e-mail: zwzhang@hrbeu.edu.cn; Zhongwu_zhang@hotmail.com

G. Lambotte et al. (eds.), *Materials Processing Fundamentals 2018*,
The Minerals, Metals & Materials Series,
https://doi.org/10.1007/978-3-319-72131-6_3

treatment. Different aging time and temperature have a significant effect on size and number density of nanophase precipitates, and finally mechanical properties [8–10].

Recently, the precipitation of nanoscale Cu-enriched precipitates in high strength steels has attracted considerable attentions, which give an effective way to strengthen the steels without significantly sacrifice ductility [11, 12]. According to the precipitation mechanism, it is generally considered that coherent body-centered cubic (bcc) Cu nanoparticles nucleate from the supersaturated solid solution, then Ni, Mn and Al were tended to segregate at the interface between the Cu nanoparticles and matrix [13–15]. Because the Cu has a limited solubility in ferritic matrix, the nanoscale Cu-rich precipitates with high number density could be obtained after aging treatment [16]. In 1930s, researchers have found when heat treatments were conducted at the range of 450–600 °C, it could lead to the precipitation of Cu element, and the yield strength could be increased by 100–200 MPa [17–19]. It is also indicated that when the precipitates with the size of 2.5–3.0 nm, the precipitation hardening peak could be obtained [20]. Some research also found that optimizing the content of alloy elements and aging processing, a proper Cu-rich precipitates and matrix microstructure can be obtained, leading to excellent mechanical properties [21]. Meanwhile, excellent plasticity and fracture toughness strongly depend on the interaction of precipitates and dislocations [22, 23]. With changing the treated temperature and time, the dislocation configuration, dislocation density and the movement of dislocation would be powerfully affected. Therefore, optimizing aging process by controlling aging time and temperature is the best way to control the mechanical properties of nanoprecipitates-strengthened ferritic steels.

The purposes of this work are to study the effect of aging treatment on the microstructure, phase composition and mechanical properties in nanoprecipitates strengthened steels. The size and number density of nanophase are regulated by reasonably distribute aging time and temperature. Mechanical behaviors of samples with different aging treatment are compered, and the influence of aging process on properties is analyzed.

Experimental

The nanoprecipitates-strengthened ferritic steel was melted in a vacuum-induction furnace under an argon atmosphere. The chemical composition of sample was list in Table 1. The ingot was homogenized at 1100 °C for 2 h and hot rolled to 12 mm over a series of steps at 900 °C. The rolled plate was quenched to room temperature. The as-rolled samples were aged at 500, 550 and 600 °C, respectively from 30 min to 50 h, followed by water quench.

The Vickers hardness measurement was performed under a load of 1000 g, and every sample was test over 10 random locations. The tensile samples with 5 mm in diameter and 25.4 in gauge length were prepared by numerically controlled lathe. And the tensile tests were conducted on INSTRON 5565 at room temperature with

Table 1 Chemical composition of steel (wt.%)

C	Si	P	S	Mn	Mo	Cr
<0.05	0.20	<0.004	<0.003	2.00	0.5	0.5
Cu	Ni	Al	B	Nb	Ti	Fe
2.50	2.00	1	0.002	0.03	0.05	bal

a strain rate of ~1 × 10-3/s. The yield strength was determined with the 0.2% off set plastic strain method.

The phase components of the specimens were determined by X-ray diffraction (XRD): Cu Target K alpha ray, tube pressure is 40 kV tube pressure, 20°–90° scanning angle, and 5°/min of scanning speed. Optical microscopy (OM) and scanning electron microscope (SEM) were also used to characterize the microstructure of the specimens. The OM and SEM samples were polished to a final surface finish of 0.2 μm by standard mechanical polishing procedures. Then the samples were etched with a 4 vol.% nital solution.

Results and Discussion

Effect of Aging Treatment on Microstructure

The microstructures of the samples aged at 500 °C for different time are shown in Fig. 1. After aging treatment, the microstructure of the samples was polygonal ferrite, and there was not obvious second phase in the matrix. The microstructure of matrix remained with increasing the aging time from 1 to 50 h. However, the grain size of samples was increased with prolonging aging time. When the sample was aged at 500 °C for 1 h, the average of grain size was about 10 μm. Improving the aging time to 2 and 5 h, the grain size slightly increased to 15 and 20 μm, respectively. Keeping the holding time to 20 and 50 h, the grain size was similar (about 30 μm). Therefore, when the nano-precipitates-strengthened ferritic steels were aged at 500 °C, the microstructure was no affected by aging time, and the grain size was increased with prolonging treating time. And when the aging time above 20 h, the grain size basically remained.

Figures 2 and 3 shows the microstructures of the samples aged at 550 and 600 °C for 1, 2, 5, 10, 20 and 50 h, respectively. When increasing the aging temperature to 550 and 600 °C, the microstructures remained polygonal ferrite, which is same to that of the samples treated at 500 °C. For the samples treated at 550 °C for 5 h, the grain size improved to 25 μm and that is same to the samples treated at 600 °C for 2 h. Therefore, the grain size also increased with aging time, and the grow rate was enhanced with increasing the aging temperature. It can be concluded that aging temperature and aging time do not change the microstructures significantly.

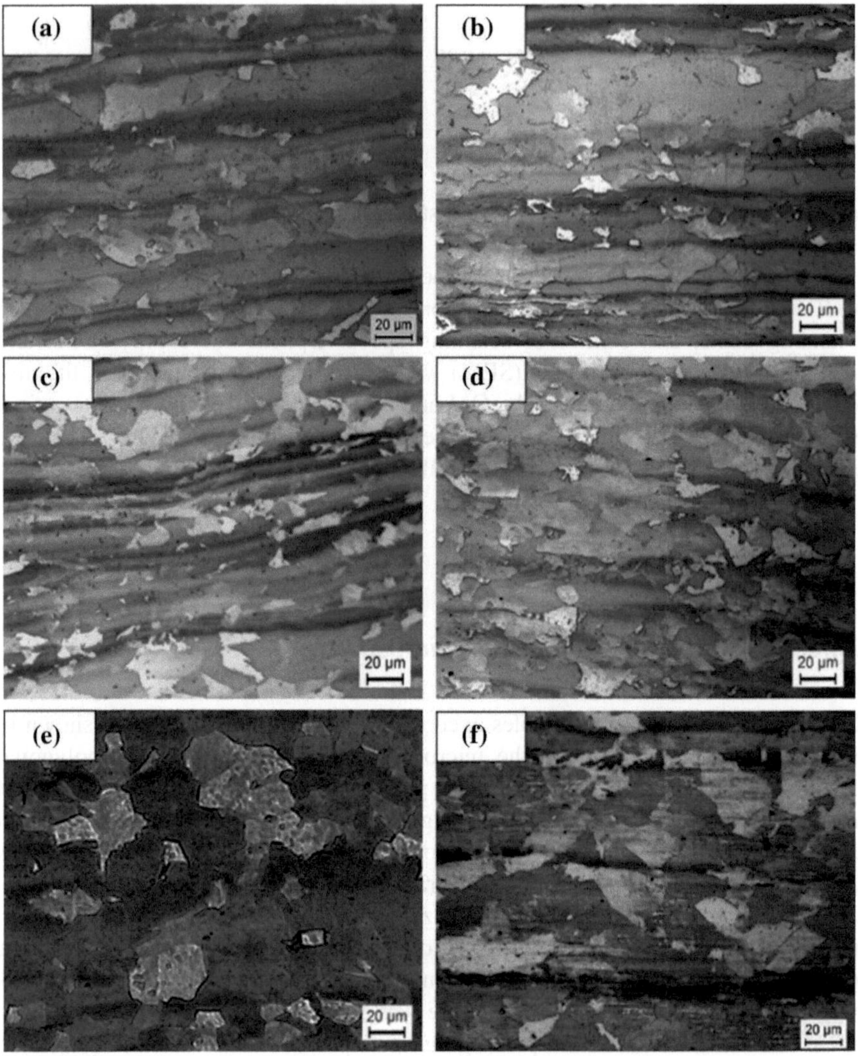

Fig. 1 Microstructures of the samples aged at 500 °C for **a** 1 h, **b** 2 h, **c** 5 h, **d** 10 h, **e** 20 h and **f** 50 h

Effect of Aging Treatment on Phase Composition

Figure 4 shows the diffraction spectra of this steel with different aging treatments. It can be seen from XRD patterns exhibited three peaks, which corresponded to the (110), (200) and (211) crystallographic planes, respectively. When the aging

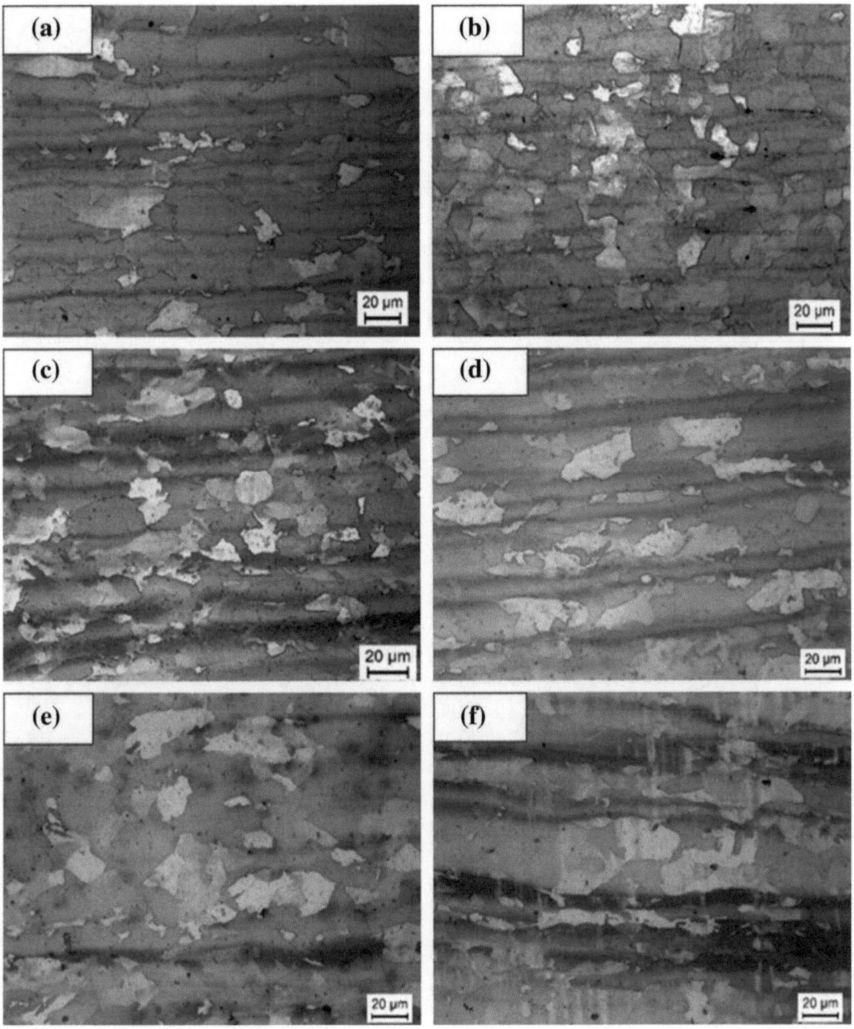

Fig. 2 Microstructures of the samples aged at 550 °C for **a** 1 h, **b** 2 h, **c** 5 h, **d** 10 h, **e** 20 h and **f** 50 h

temperature and aging time were increased, the XRD patterns were the same, there was ferrites in the matrix and no another phase was found. It is indicated that phase composition was not effect by aging temperature and aging time.

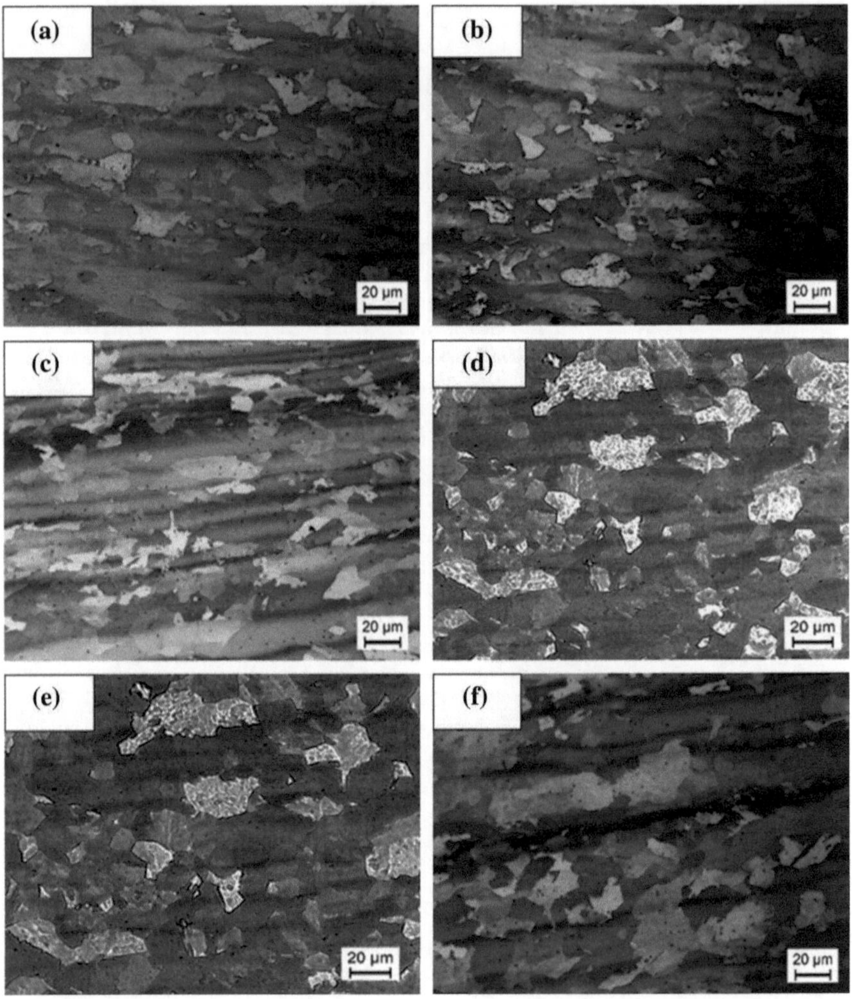

Fig. 3 Microstructures of the samples aged at 600 °C for **a** 1 h, **b** 2 h, **c** 5 h, **d** 10 h, **e** 20 h and **f** 50 h

Effect of Aging Treatment on the Mechanical Properties

The results of microhardness testing are shown in Fig. 5. The microhardness of all samples aged at different temperatures has three stages: under-aging, peak-aging, and over aging. For samples aged at 500 °C, the aging peak present at 2 h and all the hardness were higher than that of as-rolled samples (~280 HV). With prolonging the aging time at different temperature, the hardness increases first and then decreases. The presence of nanophase is the mainly result of improving hardness. When further increasing aging time, the hardness remained stable. For the samples

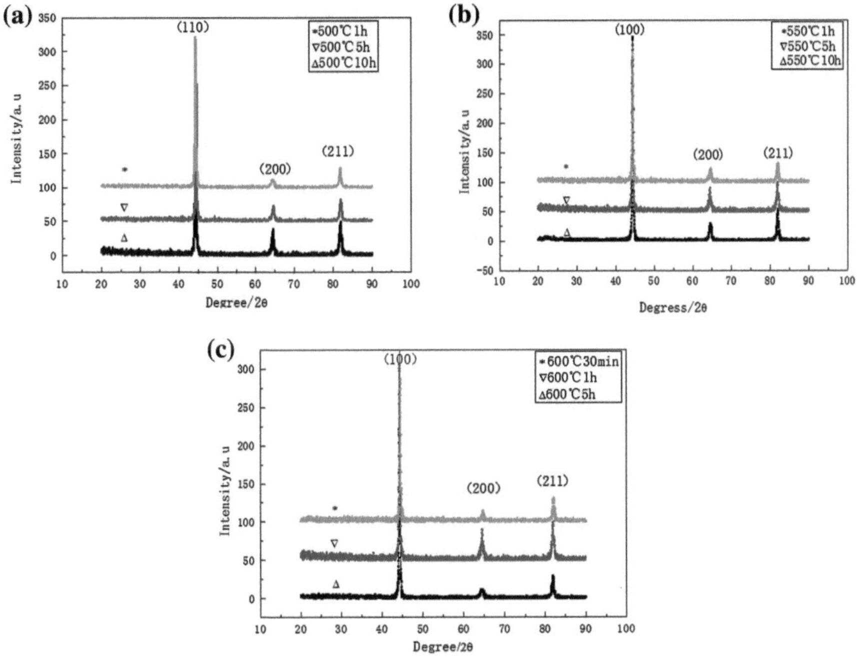

Fig. 4 X-ray diffractograms of steel with different aging treatment: **a** at 500 °C for 1, 5 and 10 h; **b** at 550 °C for 1, 5 and 10 h; **c** at 600 °C for 30 min, 1 h and 5 h

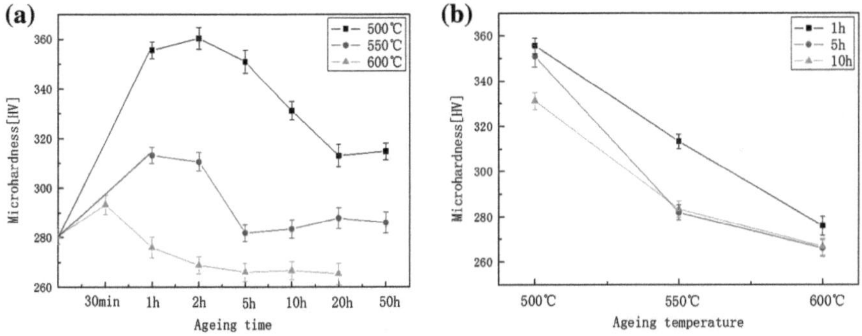

Fig. 5 The Vickers microhardness as a function of: **a** aging time and **b** aging temperature

aged at 550 °C, the peak of hardness was taken forward to 1 h, then the HVN was reduced clearly. Further increasing the aging temperature to 600 °C, the peak of hardness was 30 min. From the Fig. 5b, the hardness would be decreased with increasing the aging temperature. Therefore, in this steel, the microhardness could be effectively control by aging treatment.

The effect of aging time on tensile properties was shown in Fig. 6. When the steel was aged at 500 °C, the strength was decreased with increasing aging time, and the ductility was enhanced. For the samples aged at 500 °C for 1, 5 and 10 h, the tensile stress was 1159.8, 1106.4, and 1022 MPa, respectively. Then the yield stress was 1135.5, 1065.5 and 989.1 MPa. And the elongation was 11.9, 12.3 and 15.3%. It can be founded that the variation tendency of the strength with aging time was like to the tendency of microhardness versus aging time. Different from the samples aged at 500 °C, the yield stress of samples aged at 550 °C was observably declined with aging time. The yield strength of the sample aging 10 h brings down to 895.9 MPa and the elongation was increased slightly. When increasing the aging temperature to 600 °C, the strength decreased further, but the ductility was remained. Therefore, it can be indicated that the strength of samples decreased and elongation slightly increased with increasing aging time under the same aging temperature.

The effect of aging temperature on tensile properties was shown in Fig. 7. When the samples were aged for 1 h under different aging temperature, we can find that the yield stress will declined and elongation will improve with adding aging temperature. Further increasing aging time to 5 h, the descend range of strength was increased, and the extension rate does not change much. It is known that the aging

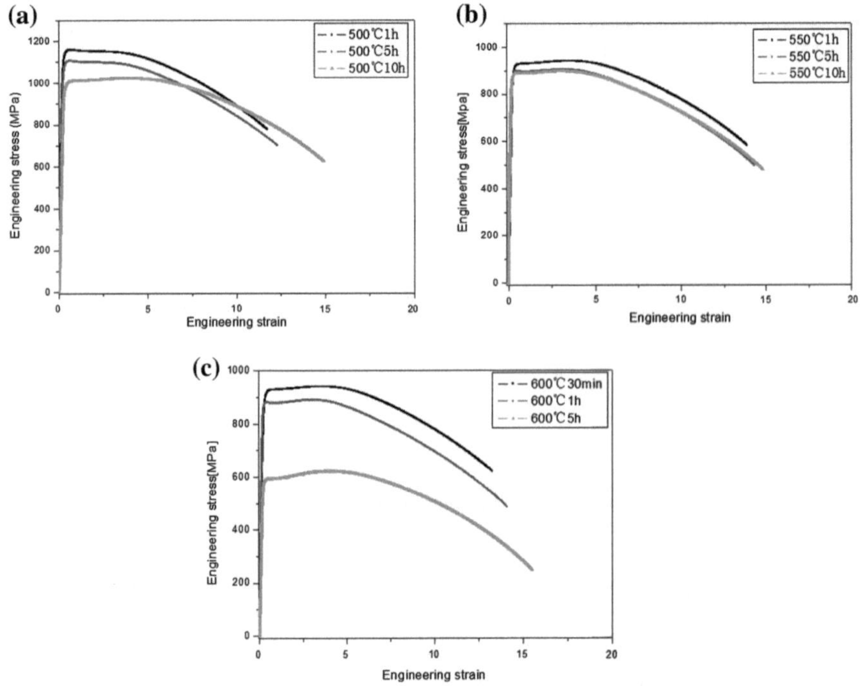

Fig. 6 Engineering stress–strain curves of the samples treated at **a** 500 °C, **b** 550 °C and **c** 600 °C for different aging time

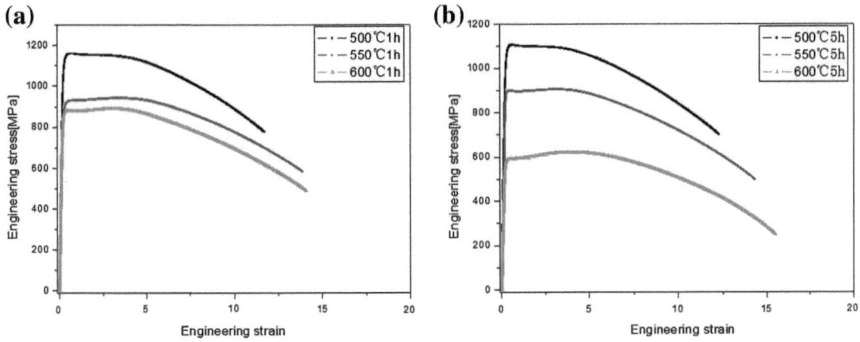

Fig. 7 Engineering stress–strain curves of the samples treated at different aging temperatures for **a** 1 h and **b** 5 h

process leads to the formation of nanoscale Cu-rich precipitates, which can enhance the strength significantly without indistinctively sacrificing elongation. It is because that the interaction of dislocations improved the strength of steel, while due to the core of the nanophase was Cu cluster, and the dislocations can cut through the nanophase, So the strength increased without sacrificing the elongation.

Comprehensive analysis above results, the aging treatment has an important influent on mechanical properties of nanoprecipitates strengthened steels. Therefore, controlling the aging temperature and aging time, which could effectively control the nucleation and grown of nanophase. That means optimized aging treatment can obtain the best mechanical properties.

SEM Analysis of Tensile Fracture

The fracture surfaces of the samples with different aging treatment were shown in Fig. 8. For the sample aged at 500 °C for 1 h (Fig. 8a), the fracture morphology was ductile fracture, and lots of small dimples could be seen. Then enlarge the aging temperature to 10 h (Fig. 8b), the fracture pastern was same to that of samples aged at 500 °C for 1 h, but the size of dimples was improved. When increasing the temperature to 550 and 600 °C, the fracture morphology remains the same, and the size of dimples was improved with temperature. It can be found that all the fracture surfaces of the samples were belong to plastic fracture. Increasing aging temperature and time, the fracture mechanisms remained. These cavities undergo nucleating, growing up and then gather to the fracture under extra loading. Therefore, this steel kept microvoid coalescence fracture, which belongs to plastic fracture. And that also proves that the present of nanoprecipitates will not sacrifice the ductility of this steel.

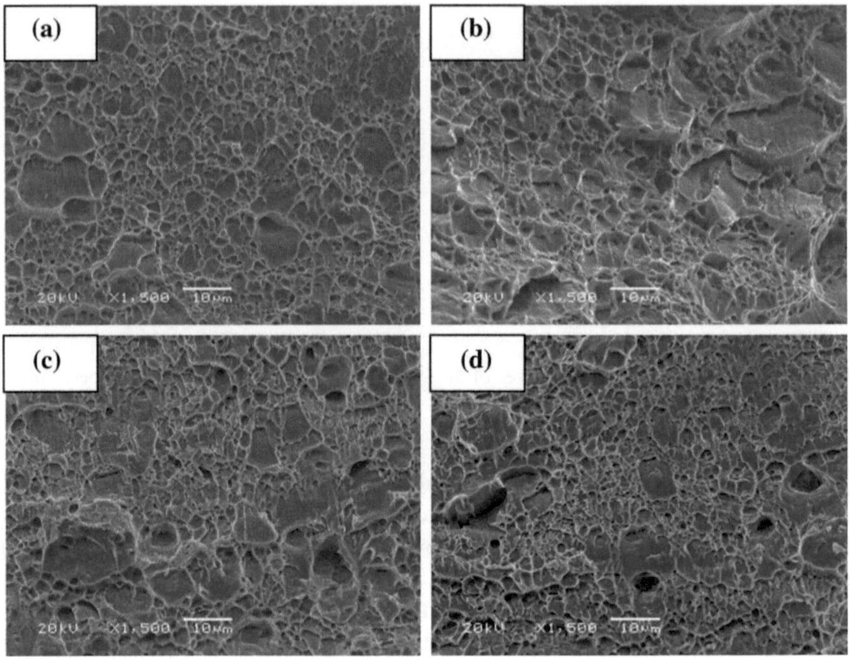

Fig. 8 Tensile fracture morphologies of the samples with different aging treatments **a** 500 °C for 1 h, **b** 500 °C for 10 h, **c** 550 °C for 1 h and **d** 600 °C for 1 h

Conclusions

Nano-precipitates strengthened steels as an important structural material, and the main strengthened mode is precipitation strengthening. In this study, the effect of aging temperature and aging time on mechanical properties was detail investigated. The following conclusions can be drawn:

1. The aging temperature and aging time no significantly effect on microstructure, and the microstructure of matrix remain polygonal ferrite. Increasing the aging temperature and aging time do not cause the change of microstructure, and just adding the size of ferrites. While, the phase composition was not influence on aging treatment, changing aging time and aging temperature, the phase of matrix keep ferrite.
2. Aging treatment can effectively influence the microhardness. With increasing aging time, the hardness presents the stages of under-aging, peak-aging, and over aging. Then the aging peak of hardness will move to an earlier with improving aging temperature.
3. The changing trend of strength is like the variation of hardness. The highest strength present at the aging peak. Under same aging time, the tensile stress and yield stress decreased with the increase of aging temperature. The optimal

matching of strength and ductility was obtained at 500 °C for 10 h, and the strength was 1022 MPa, elongation was 15.3%.
4. All fracture mechanisms of aging samples were microporous aggregation fracture. Then the size of dimples increased with improving aging time and temperature.

Acknowledgements This work was supported by the Fundamental Research Funds for the Central Universities (HEUCFJ171005, HEUCFJ171001 and HEUCFP201703), NSFC Funding (51371062 and U1460102), NSFHLJ (JC2017012), the Scientific Research Foundation for Returned Overseas Chinese Scholars (Heilongjiang Province), the High-Tech Ship Research Projects Sponsored by MIIT (K24367). Yu Zhao is benefited from the International Exchange Program of Harbin Engineering University for Innovation-oriented Talents Cultivation.

References

1. Zhang ZW, Liu CT, Wen YR, Hirata A, Guo S, Chen G, Chen MW, Chin BA (2011) Metall Mater Trans A 43:351–359
2. Jiao ZB, Luan JH, Zhang ZW, Miller MK, Ma WB, Liu CT (2013) Acta Mater 61:5996–6005
3. Jiao ZB, Luan JH, Miller MK, Liu CT (2015) Acta Mater 97:58–67
4. Ghosh A, Sahoo S, Ghosh M, Ghosh R, Chakrabarti D (2014) Mater Sci Eng A 613:37–47
5. Xu W, Rivera-Díaz-Del-Castillo PEJ, Yan W, Yang K, Martín DS, Kestens LAI, Zwaag SVD (2010) Acta Mater 58:4067–4075
6. Zhang ZW, Liu CT, Guo S, Cheng JL, Chen G, Fujita T, Chen MW, Chung Y-W, Vaynman S, Fine ME, Chin BA (2011) Mater Sci Eng A 528:855–859
7. Vaynman S, Isheim D, Kolli RP, Bhat SP, Seidman DN, Fine ME (2008) Metall Mater Trans A 39:363
8. Kapoor M, Isheim D, Vaynman S, Fine ME, Chung YW (2016) Acta Mater 104:166–171
9. Fitzgerald SP, Dudarev SL (2009) J Nucl Mater 386–388:67–70
10. Jain D, Isheim D, Hunter AH, Seidman DN (2016) Metall Mater Trans A 47:1–13
11. Zhang ZW, Liu CT, Miller MK, Wang XL, Wen Y, Fujita T, Hirata A, Chen M, Chen G, Chin BA (2013) Sci Rep 3:975–978
12. Harry T, Bacon DJ (2011) MRS Online Proceeding Library, 439
13. Dhua SK, Mukerjee D, Sarma DS (2003) Metall Mater Trans 34A:241
14. Zhao Y, Xu SS, Li JP, Zhang J, Sun LW, Chen L, Sun GA, Peng SM, Zhang ZW (2017) Mater Sci Eng A 691:162–167
15. Mishra SK, Das S, Ranganathan S (2002) Mater Sci Eng A 323:285–292
16. Misra RDK, Venkatsurya P, Wu KM, Karjalainen LP (2013) Mater Sci Eng A 560:693–699
17. Fine ME, Isheim D (2005) Scripta Mater 53:115
18. Rashid MS (1980) Science 208:862–869
19. Isheim D, Gagliano MS, Fine ME, Seidman DN (2006) Acta Mater 54:841
20. Ray PK, Ganguly RI, Panda AK (2003) Transactions of the Indian Institute of Metals
21. Movahed P, Kolahgar S, Marashi SPH, Pouranvari M, Parvin N (2009) Mater Sci Eng A 518:1–6
22. Suarez MA, Alvarez-Pérez MA, Alvarez-Fregoso O, Juarez-Islas JA (2011) Mater Sci Eng A 528:4924–4926
23. Wen YR, Hirata A, Zhang ZW, Fujita T, Liu CT, Jiang JH, Chen MW (2013) Acta Mater 61:2133–2147

Part III
Multiphysics—Process Modeling and Sensing

Convection-Diffusion Model of Lithium-Bismuth Liquid Metal Batteries

Rakan F. Ashour and Douglas H. Kelley

Abstract The liquid metal battery is composed of two liquid metals with different electronegativity separated by molten salt. The three layers self-segregate based on density allowing for easy manufacturing and scaling. Lithium (Li) is one of the most widely researched anode materials, and when coupled with bismuth (Bi) cathodes, it gives a liquid metal battery that has an open circuit voltage of 0.9 V. Such a system has demonstrated impressive rate capabilities, ultra-long life cycle, and low energy cost. Here we present a two-dimensional physics-based model for Lithium-Bismuth liquid metal batteries. The model takes into account dynamical changes in the battery, including surface concentration and fluid flow. By solving the convection-diffusion equation in Bi electrodes, we are able to investigate the effect fluid flow on kinetic losses and concentration profiles in real time. The outcome of this work allows us to link electrochemistry and fluid dynamics in liquid metal batteries. Moreover, the model can also be used to guide future development of battery management systems.

Keywords Liquid metal battery · Battery model · Magnetohydrodynamics

The liquid metal battery is composed of two liquid metals with different electronegativity separated by molten salt. The three layers self-segregate based on density allowing for easy manufacturing and scaling. Lithium (Li) is one of the most widely researched anode materials, and when coupled with bismuth (Bi) cathodes, it gives a liquid metal battery that has an open circuit voltage of 0.9 V. Such a system has demonstrated impressive rate capabilities, ultra-long life cycle, and low energy cost. Here we present a two-dimensional physics-based model for Lithium-Bismuth liquid metal batteries. The model takes into account

R. F. Ashour (✉)
Materials Science Program, University of Rochester, Rochester, NY, USA
e-mail: rashour@ur.rochester.edu

D. H. Kelley
Department of Mechanical Engineering, University of Rochester, Rochester, NY, USA
e-mail: d.h.kelley@rochester.edu

G. Lambotte et al. (eds.), *Materials Processing Fundamentals 2018*,
The Minerals, Metals & Materials Series,
https://doi.org/10.1007/978-3-319-72131-6_4

dynamical changes in the battery, including surface concentration and fluid flow. By solving the convection-diffusion equation in Bi electrodes, we are able to investigate the effect fluid flow on kinetic losses and concentration profiles in real time. The outcome of this work allows us to link electrochemistry and fluid dynamics in liquid metal batteries. Moreover, the model can also be used to guide future development of battery management systems.

Introduction

Today's electrical grids were designed back when energy storage did not exist. Hence, they require constant monitoring in order to maintain the balance between energy generation and consumption by end users. The integration of renewable energy sources such as wind and solar adds to the complexity of grid management due to the fact that these sources are intermittent. Energy storage can allow the seemingly impossible task of decoupling supply from demand. Moreover, it allows electrical grids to operate at a steady base-load. According to a 2013 report by the U.S Department of Energy, the United States will require 4 to 5 TWh of storage capacity in order to accommodate the rising demand for renewable energy integration [1]. However, grid-scale storage technologies must be cost-competitive without subsidies and must be able to integrate into existing infrastructure. In fact, the US Advanced Research Projects Agency-Energy (ARPA-e) has defined cost competitiveness as the ability to deliver energy at a cost of 100$/kWh [1, 2]. One technology that meets these strict cost requirements is the liquid metal battery (LMB) [3].

LMBs are composed of two molten metals from opposite sides of the periodic table separated by a mixture of molten salts. Candidates for anode materials include alkali and alkaline metals such as Lithium (Li), [3, 4] sodium (Na), [5, 6] or calcium (Ca) [7, 8]. Candidates for cathode materials include antimony (Sb), bismuth (Bi) and zinc (Zn) [9]. Both the anode and cathode materials are selected to ensure the battery remains liquid during operation. The electrolyte mixture, typically composed of molten metal halides, hydroxides, amides or mixtures thereof, [10] is chosen in order to maximize ionic conductivity and minimize operating temperature. Since the three layers are self-segregated based on density, the assembly of these batteries does not require sophisticated manufacturing techniques, hence, lowering the overall cost of the battery. A schematic representation of typical liquid metal battery assembly is shown in Fig. 1.

Li-based LMBs are of great interest due to their high potential, low solubility in molten salts, high ionic conductivity of molten halides, [11] and ultra-long life cycle [3]. Of particular interest is the Li||Bi LMB due to its high voltage (0.9 V) and intermediate operating temperature (<400 °C). In addition, Li||Bi batteries allow deep cycling to the point where the cathode undergoes a phase transformation from liquid Bi to a completely solid Li–Bi intermetallic. Hence, it is the combination of

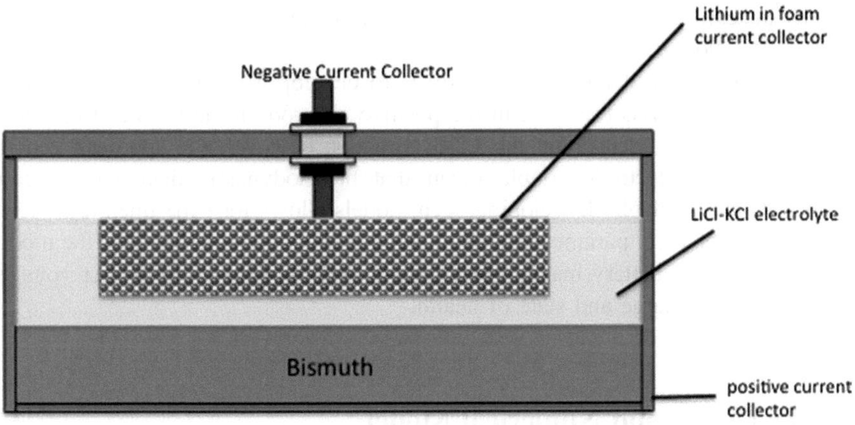

Fig. 1 Schematic representation of a lithium based liquid metal battery. The model focuses on transport in the positive (bismuth) electrode

these exciting features, the availability of thermodynamic data [12] and material properties [13] that attracted our attention this system.

When the Li||Bi LMB is connected to an external load, Li metal ions are produced via oxidation (Li^+). These metal ions have to travel through the molten electrolyte before they are reduced at the surface of the positive electrode where they form an alloy ($Li_{in\text{-}Bi}$). The open circuit potential of the battery depends on the Gibbs free energy of the alloying reaction. Upon discharge, N moles of Li metal oxidize to produce a current (I). Products are then carried by convection and diffusion from the interface to the bulk of the liquid electrode. The accumulation of Li in Bi or the formation of intermetallic at the electrode-electrolyte interface causes a substantial drop in the potential of the battery due to the rise of transport over-potential. Previous experimental work by Kim et al. observed a significant decrease in capacity at high current densities due to the formation of a two-phase mixture in the positive electrode of Ca||Bi battery [14]. The electrochemical measurements performed therein attribute the potential drop at high discharge current densities to the slow transport of Ca^+ in the low-density $Ca_{11}Bi_{10}$ intermetallic forming at the electrode-electrolyte interface. Li based LMBs are subject to the formation of solid intermetallic near the electrode/electrolyte interface. Interestingly however, the diffusivity of Li^+ ions is found to be higher in the solid LiBi, and Li_3Bi intermetallic, than liquid Bi [13]. However, since diffusion is not the sole mechanism that drives mass transport, especially in large-scale batteries, it is important to understand how the presence of fluid flow affects the overall kinetics. Additionally, knowledge of how concentration gradients are affected by fluid flow is important in order to anticipate the growth of intermetallic and prevent the risk of electrical shorts. If intermetallic grows in a manner that allows the two electrodes to come in direct contact, then they will react chemically and electrons are exchanged locally instead of being released to an external circuit, hence, the battery becomes electrically shorted.

The main focus of the present study is to develop a magnetohydrodynamics-based model for mass transport in the positive electrode in liquid metal batteries. The present study focuses on the Li||Bi systems; however, the approach can be applied to any electrode couple given that thermodynamic data and material properties are available. Developing such models allows for experimenting with a wide range of battery parameters with minimal cost and time. Ultimately, the model can evolve into a battery management system that can be used to predict voltage, power, state of charge and state of health.

Mathematical and Numerical Model

To generate a mathematical model that describes transport in the positive electrode, we have to consider the forces acting on the electrode and understand how they generate motion. One force that can initiate mass transport is electromagnetic (Lorentz) force, which is generated from the interaction of charge or discharge current density vector **J** with its associated magnetic field **B**. The electromagnetic force can be determined by first solving Laplace's equation for the potential distribution in the electrode:

$$\nabla^2 V = 0 \tag{1}$$

where ∇ is the Laplace operator and V is the electrical potential distribution. The resulting potential distribution is used to calculate the electric field (**E**) and current density from Eqs. (2) and (3) respectively.

$$\boldsymbol{E} = \nabla \boldsymbol{V} \tag{2}$$

$$\boldsymbol{J} = \sigma \boldsymbol{E} + \boldsymbol{u} \times \boldsymbol{B} \tag{3}$$

where σ is the electrical conductivity and **u** is the velocity vector. Here we consider the case where the induced current (**u** × **B**) is negligible compared to the applied Current ($\sigma\boldsymbol{E}$). The magnetic field is then determined from Ampere's law and the resulting electromagnetic force per unit volume is the cross product of **J** and **B** (Eqs. 4 and 5 respectively)

$$\oint \boldsymbol{B} d\boldsymbol{l} = \mu_o \oiint \boldsymbol{J} d\boldsymbol{A} \tag{4}$$

$$\boldsymbol{F} = \mathbf{J} \times \boldsymbol{B} \tag{5}$$

where μ_o is magnetic permeability constant. In a typical liquid metal battery with cylindrical geometry, the current density vector **J** is symmetric about the central axis with components in the radial ($\mathbf{J}_r$) and axial component ($\mathbf{J}_z$). This will generate an azimuthal magnetic field $\mathbf{B}_\theta$ and in the absence of thermal convection will tend to drive an axisymmetric flow that sinks in the center and rises near the sidewalls [15]. Note that switching from charge to discharge does not change the direction of the electromagnetic force.

Another force that can initiate mass transport is thermal buoyancy. Buoyant forces arise whenever temperature gradients are generated in a body of fluid: a recirculating flow is created in which gravitational energy is converted into kinetic energy. However, these forces are also opposed by viscosity, and in the case where viscous forces are strong enough, motion can be completely suppressed. The onset at which buoyancy overcomes viscosity is predicted by a dimensionless quantity that represents their relative ratio as

$$Ra = \frac{g\beta \Delta T L^3}{v\alpha} \tag{6}$$

where β is the coefficient of thermal expansion, g is the gravitational acceleration, ΔT is the temperature difference, L is the thickness of the electrode, ν is the kinematic viscosity and α is the thermal diffusivity. This dimensionless quantity is called the Rayleigh number and the transition from viscous to buoyancy-dominated regime is called the Rayleigh-Bénard instability [15]. For geometries relevant to liquid metal batteries with one free surface, the Rayleigh-Bénard instability occurs at a critical value of $Ra_c \sim 1100$, though the exact value depends on geometric details of the fluid vessel [15].

Typical liquid metal batteries require heating a positive current collector from the bottom. Consequently, temperature gradients are generated along the depth of each fluid layer. Moreover, with the use of imperfect thermal insulation, we expect the bottom fluid (cathode) to be hotter near the center and colder near the edges. Therefore, we expect thermal buoyancy to drive a flow in which fluid elements rise near the central axis and sink near the edges [16]. Combining all terms, momentum transfer is governed by the Navier-Stokes equation, which is the application of Newton's second law to viscous fluid systems,

$$\frac{\partial \mathbf{u}}{\partial t} + (\mathbf{u} \cdot \boldsymbol{\nabla})\mathbf{u} = \boldsymbol{v}\boldsymbol{\nabla}^2\mathbf{u} - \frac{1}{\boldsymbol{\rho}}\boldsymbol{\nabla}\boldsymbol{p} + \frac{1}{\boldsymbol{\rho}}(\boldsymbol{J} \times \boldsymbol{B}) - \boldsymbol{\alpha}\boldsymbol{\Delta}\boldsymbol{T}\boldsymbol{g}\hat{\boldsymbol{z}} \tag{7}$$

where $\frac{\partial}{\partial t}$ is the time derivative, p is pressure, ρ is the density of the electrode, and $\hat{z}$ is a unit vector in the vertical direction. Note that buoyancy driven flow would be moving in an opposite direction to that driven by electromagnetic forces. Hence, one would naturally ask: will buoyant or electromagnetic forces dominate the flow? Current experimental and theoretical work is undertaken by our group to answer this question. However, for the purpose of our model, we shall assume that electromagnetic forces are negligible and do not affect the overall flow structure. This assumption is supported by the theoretical work of Ejik et al. dealing with flow transitions in vacuum arc re-melting, [17] as well as the mathematical models developed by Davidson [15]. In fact, both references state that in the ideal case where temperature gradients are fixed with increasing current density, there exists a wide range of current densities at which the electromagnetic force seems to have no effect on the flow structure.

So far our discussion has focused on building a physical intuition of the forces that can initiate flow. Let us now consider how we can link chemical concentration with fluid flow. Changes in chemical concentration are described by the convection-diffusion equation as

$$\frac{\partial c}{\partial t} = \nabla \cdot (\boldsymbol{D} \nabla \boldsymbol{c}) - \nabla \cdot (\boldsymbol{uc}) \tag{8}$$

where c is the concentration profile, D is the material diffusivity, and $\boldsymbol{u}$ is the velocity vector. One approach is to sequentially solve Eqs. (7) and (8). Clearly, this is not an easy task since solving the momentum equation by itself is computationally challenging. An alternative is to use a stream function that approximates the dynamics in the core and neglects boundary layer effects. A time independent stream function (Ψ) that represents symmetric convection rolls is written as

$$\Psi \sim \sin\frac{2\pi x}{d}\sin\frac{\pi y}{L} \tag{9}$$

where x and y are horizontal and axial spatial coordinates, respectively, and d is the diameter of the positive electrode. It is important to note that a stream function will always satisfy the incompressibility condition ($\nabla \cdot \boldsymbol{u} = \boldsymbol{0}$). This will further simplify Eq. (8), and guarantees the conservation of mass within the electrode. In each simulation run, the velocity profile ($\boldsymbol{u} = U(\nabla \times \boldsymbol{\Psi})$ is considered an input in the convection-diffusion solver, where U is the characteristic flow speed. The overall flow field is shown in Fig. 2. The operating current using Faraday's law sets the flux across the interface:

$$\mathrm{C} = \frac{\mathrm{J\ dt}}{n_e F} \tag{10}$$

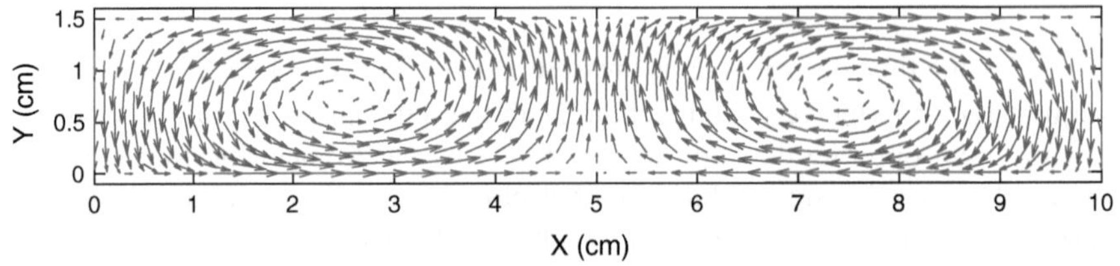

Fig. 2 Flow field specified by Eq. 9 and intended to model thermal convection inside a bismuth electrode

where C is the molar concentration (mol/cm^2), n_e is the number of electrons exchanged in the reaction, and F is Faraday's constant (96,485 C mol^{-1}). The numerical solver uses first-order finite differences for spatial derivatives and a second-order Adams-Bashforth scheme for time derivatives. The employed boundary conditions are constant flux across the cathode-electrolyte interface, which represents a battery discharged at a constant current, and no-flux across other walls.

Previously published electromotive force data [12, 13] are used to estimate the open circuit voltage at different molar fraction (X_{Li}). The molar fraction and the corresponding open circuit voltage are calculated using Eqs. 11 and 12 respectively

$$X_{Li} = \frac{C_{Li}}{C_{Li} + C_{Bi}} \quad (11)$$

$$V(X_{Li}) = \begin{cases} \frac{0.67}{X_{Li}} & for\, X_{Li} < 0.3 \\ 0.75 & for\, 0.3 \le X_{Li} < 0.74 \\ 0.05 & for\, X_{Li} > 0.74 \end{cases} \quad (12)$$

where C_{Li} and C_{Bi} are the molar concentration of Li and Bi respectively, and $V(X_{Li})$ is open circuit voltage. The operating temperature is chosen to be 400 °C, and the electrolyte layer is composed of a eutectic mixture of LiCl and KCl. The change in diffusivity with concentration is accounted for in the model, and the overall diffusivity D in cm^2/s is found in the work of Newhouse et al. [13] as

$$D = e^{\frac{(-36.57X_{Li} - 4.27)}{(X_{Li}^2 + 3.07X_{Li} + 0.5)}} \quad (13)$$

Results and Discussion

We first choose an electrode thickness such that the Rayleigh number in Eq. (6) is greater than the estimated critical value of 1100 even when temperature gradients are as low as 2 °C. For this case, we choose a thickness of 1.5 cm in a 10 cm diameter container with a 6 cm foam negative current collector similar to those observed in previously published work [4]. The phase diagram of Li–Bi predicts the onset of the two-phase region at $X_{Li} < 30\%$. Hence, knowing that a total of 5.6 mol of Bismuth are used in the positive electrode, and since there are 26.8 Ahr/mole, the total capacity for this all-liquid battery model is approximately 45 Ahr.

Since current travels through the path of least resistance, we anticipate that current density lines will spread little through the depth of the electrolyte. Therefore, we choose to set our constant flux boundary condition such that Li ions enter the Bi electrode from an area equivalent to that of the negative electrode. To investigate the effect of convection on the overall concentration profile of Li ions, we solved Eq. (8) with constant flux corresponding to a battery discharged at a current density of 200 mA/cm^2 for 10^4 s (total discharge capacity 43 Ahr). Initially, the cathode was pure Bi; at time t = 0, Li^+ began entering the cathode via convection and diffusion. Concentration profiles for characteristic flow speeds of U = 0, 0.5 and 1 mm/s are shown in Fig. 3. The range of flow speeds is chosen to match previously observed speeds in numerical simulations [18, 19] and experimental models [16]. In the case where transport is purely driven by diffusion (U = 0), the highest concentration is observed near the electrode-electrolyte interface. When a convective flow is allowed, the maximum concentration still appears near the interface, however, the gradient is now along the direction of the flow. Due to the axisymmetric structure of the flow, the region of the central axis appears to represent a mixing barrier. Furthermore, the centers of the counter rotating vortices seem to hold lower concentrations of lithium compared to the edges. This shows that motion of Li^+ ions is driven by convection along the edges of the vortices and by diffusion towards their centers.

To follow the development of the concentration profile, we fixed convection speed to 1 mm/s and plotted snapshots of Li^+ concentration at different times. Figure 4 demonstrates that ions move much more slowly towards the centers of the vortices compared to their edges. If growth of intermetallic were allowed, then one would anticipate the direction of propagation of solid fronts to be in the same

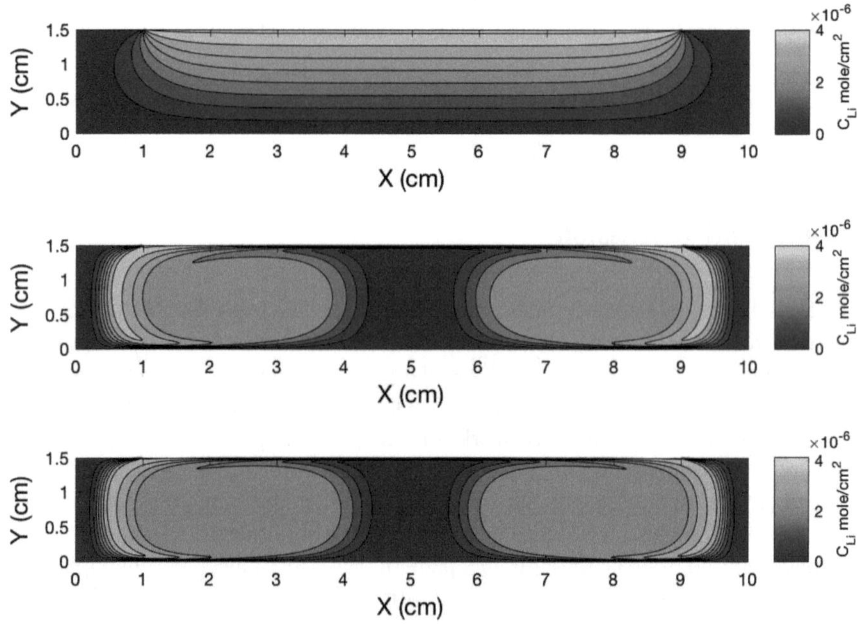

Fig. 3 Concentration of Li^+ in a Bi cathode after 10^4 s with J = 200 mA/cm^2 and different characteristic flow speeds: 0 mm/s (top), 0.5 mm/s (middle) and 1 mm/s (bottom). Thermal convection transports Li along the top surface and down the edges of the cathode, whereas diffusion spreads Li smoothly from the negative current collector

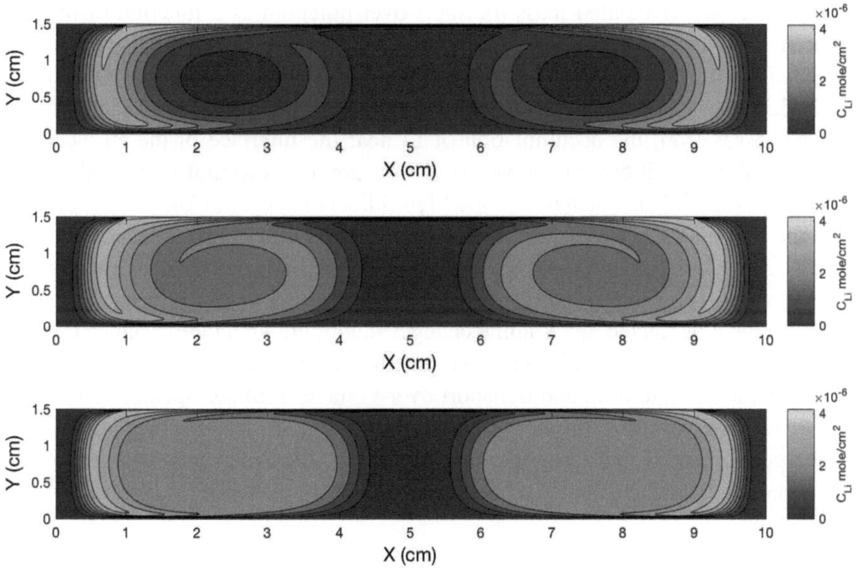

Fig. 4 Time evolution of Li^+ concentration in a Bi cathode at time t = 10^2 s (top), t = 10^3 (middle) and t = 10^4 s (bottom). In all cases, the characteristic flow speed is 1 mm/s. Li^+ is transported primarily around the vortices, via convection

direction as the concentration gradient. Consequently, intermetallic nuclei are expected to form at the interface and grow along the edges of the vortices when thermal convection is present, and uniformly towards the bottom of the electrode in the absence of convection [4].

So far we have only considered the impact of the flow on the concentration of ions. To study its impact on battery performance, we shall calculate the over-potential associated with mass transport at different flow speeds. To do this, we will use the average concentration at the interface (C^+) to calculate the surface molar fraction (X^+) of Li^+ in Bi using Eq. 11 and the surface potential $V(X^+)$ using Eq. 12. The voltage drop is then calculated by subtracting the surface potential from that of the bulk as

$$\eta_{mt} = V(X^+) - V(X_{Li}) \tag{14}$$

where η_{mt} is the voltage drop due to the accumulation of products at the electrode-electrolyte interface. The above expression takes into account local changes in the chemical activity of Li. Another possible way to calculate the over potential is to use the Nernst equation and assume that activity does not change significantly. However, since EMF measurements are available, it is more accurate to use the expression in Eq. (12). This approach was also used by Newhouse to calculate transport over-potentials in both Li||Bi and Mg||Sb systems [13].

In Fig. 5, we calculated mass transport over-potential as a function of time for different characteristic flow speeds. The rapid drop in over-potential with time is expected since ions initially accumulate at the interface and then get swept away by either diffusion or convection. Introducing fluid flow is found to lower the overall voltage losses from the accumulation of Li near the interface of the Bi electrode. However, the overall changes in voltage losses are low, especially as the discharge time increases. This is attributed to the high diffusivity of Li in Bi. To compare η_{mt} with the voltage drop due to the resistance of the electrolyte, an ionic conductivity of 1.27 S/cm, and an inter-electrode distance of 0.7 cm are used to calculate an overall ohmic drop of 0.12 V at a discharge current density of 200 mA/cm^2 [20]. Here, we have neglected any volume changes that might affect the distance between the two electrodes. The final step to estimate the discharge voltage of the battery is to subtract the ohmic drop and transport over-voltage from the open circuit voltage in Eq. 12. The resulting discharge voltage profile is shown in Fig. 6. The initial discharge voltage is only slightly higher for the electrode with a 2 mm/s flow compared to that where transport is purely driven by diffusion.

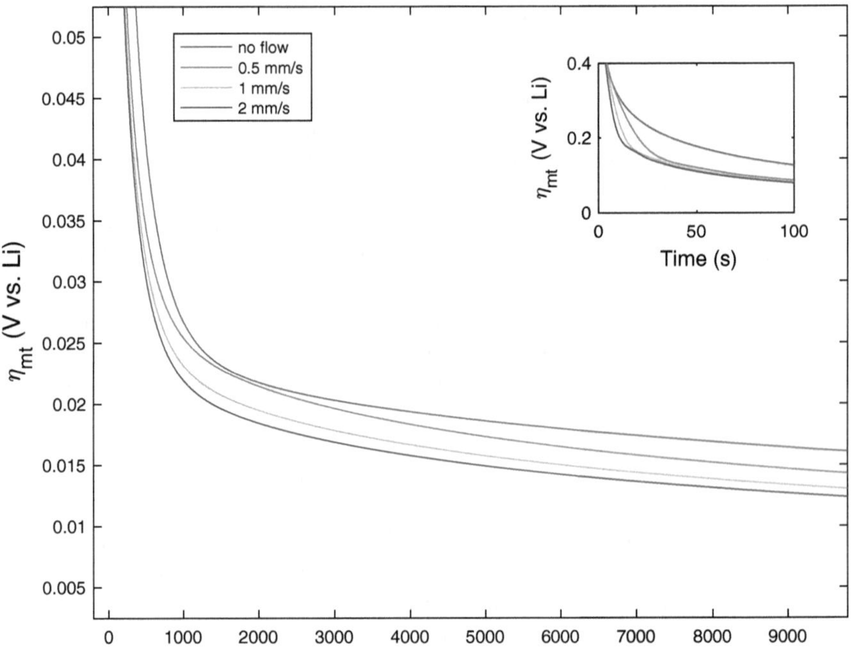

Fig. 5 Transport over-potential varying over time, at different characteristic flow speeds. J = 200 mA/cm^2 in all cases, and the inset shows rapid over-potential drop during the first few seconds after discharge begins. Higher characteristic flow speeds reduce transport over-potential

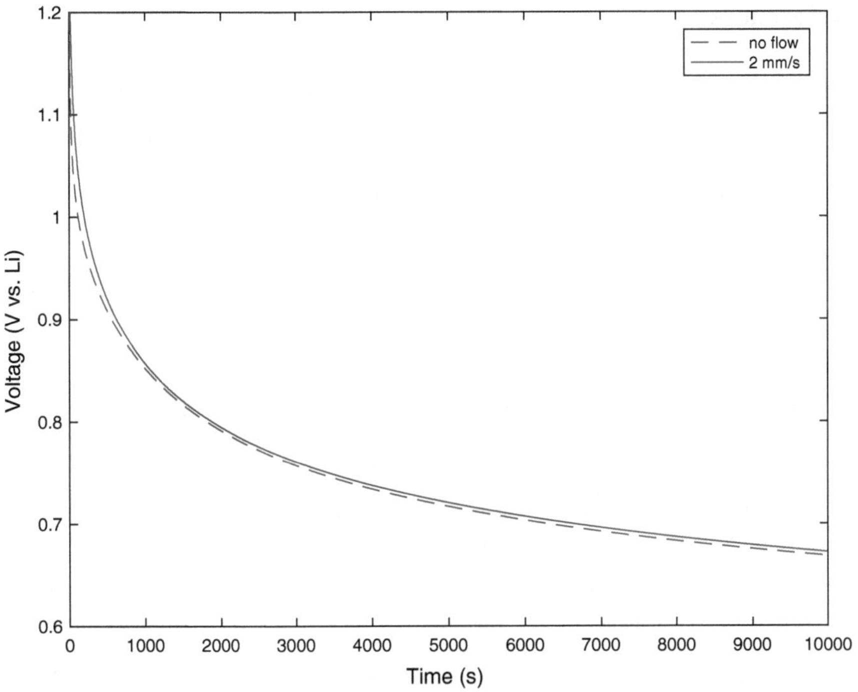

Fig. 6 Estimated discharge voltage for Li||LiCl,KCl||Bi with transport in the cathode driven by diffusion (dashed line) and 2 mm/s flow (solid blue)

Conclusion

A two-dimensional convection-diffusion model was developed to link fluid flow to battery electrochemistry. Using a simplified stream function to represent thermal convection, we studied the development of concentration profile at different flow speeds and calculated the overall kinetic losses from mass transport. Increasing the characteristic flow speed caused a slight decrease in transport over-potential. However, due to the high diffusivity of Li in Bi, the overall voltage drop due to mass transport was found to be insignificant. Therefore, our findings suggest that at parameters typical for liquid metal batteries, Li||Bi systems are kinetically limited by ohmic losses in the electrolyte and not mass transport in the cathode. Future research will focus on comparing experimental discharge voltage profiles with model predictions as well as exploring the impact of flow structure and speed on the growth of intermetallic.

References

1. U.S. Department of Energy, Grid Energy Storage, 2013. http://energy.gov/sites/prod/files/2014/09/f18/GridEnergyStorageDecember2013.pdf
2. Kim H, Boysen DA, Newhouse JM, Spatocco BL, Chung B, Burke PJ et al (2013) Liquid metal batteries: past, present, and future. Chem Rev:2075–2099
3. Wang K, Jiang K, Chung B, Ouchi T, Burke PJ, Boysen DA et al (2014) Lithium–antimony–lead liquid metal battery for grid-level energy storage. Nature. 514:348–350. https://doi.org/10.1038/nature13700
4. Ning X, Phadke S, Chung B, Yin H, Burke P, Sadoway DR (2015) Self-healing Li–Bi liquid metal battery for grid-scale energy storage. J Power Sources 275:370–376. https://doi.org/10.1016/j.jpowsour.2014.10.173
5. Ashour RF, Yin H, Ouchi T, Kelley DH, Sadoway DR (2017) Molten amide-hydroxide-iodide electrolyte for a low-temperature sodium-based liquid metal battery. J Electrochem Soc 164:A535–A537. https://doi.org/10.1149/2.1451702jes
6. Spatocco BL, Ouchi T, Lambotte G, Burke PJ, Sadoway DR (2015) Low-temperature molten salt electrolytes for membrane-free sodium metal batteries. J Electrochem Soc 162:A2729–A2736. https://doi.org/10.1149/2.0441514jes
7. Ouchi T, Kim H, Spatocco BL, Sadoway DR (2016) Calcium-based multi-element chemistry for grid-scale electrochemical energy storage. Nat Commun 7:1–5. https://doi.org/10.1038/ncomms10999
8. Kim H, Boysen DA, Ouchi T, Sadoway DR (2013) Calcium-bismuth electrodes for large-scale energy storage (liquid metal batteries). J. Power Sources 241:239–248. https://doi.org/10.1016/j.jpowsour.2013.04.052
9. Xu J, Sigmund O, Sende K, Maria A, Edvard O, Martin G (2016) Na-Zn liquid metal battery. J. Power Sources. 332:274–280. https://doi.org/10.1016/j.jpowsour.2016.09.125
10. Weier T, Bund A, El-Mofid W, Horstmann GM, Lalau C-C, Landgraf S et al (2017) Liquid metal batteries—materials selection and fluid dynamics. Mater Sci Eng 228
11. Masset P, Schoeffert S, Poinso J-Y, Poignet J-C (2005) LiF-LiCl-LiI vs. LiF-LiBr-KBr as molten salt electrolyte in thermal batteries. J Electrochem Soc 152:A405–A410. https://doi.org/10.1149/1.1850861
12. Weppner W, Huggins RA (1978) Thermodynamic properties of the intermetallic systems lithium-antimony and lithium-bismuth. J Electrochem Soc 125:7–14. https://doi.org/10.1149/1.2131401
13. Newhouse JM (2014) Modeling the operating voltage of liquid metal battery cells by, Massachusetts Institute of Technology
14. Newhouse JM, Poizeau S, Kim H, Spatocco BL, Sadoway DR (2013) Thermodynamic properties of calcium-magnesium alloys determined by emf measurements. Electrochim Acta 91:293–301. https://doi.org/10.1016/j.electacta.2012.11.063
15. Davidson PA (2001) An introduction to magnetohydrodynamics. Cambridge University Press, Cambridge
16. Kelley DH, Sadoway DR (2014) Mixing in a liquid metal electrode. Phys Fluids 26 https://doi.org/10.1063/1.4875815
17. Davidson PA, He X, Lowe AJ (2000) Flow transitions in vacuum arc remelting. Mater Sci Technol 16:55–64. https://doi.org/10.1179/026708300773002663
18. Shen Y, Zikanov O (2016) Thermal convection in a liquid metal battery. Theor Comput Fluid Dyn 30:275–294. https://doi.org/10.1007/s00162-015-0378-1
19. Thomas K, Boeck T (2016) Thermal Rayleigh-Marangoni convection in a three-layer liquid-metal-battery model, pp 1–26
20. Masset P, Guidotti RA (2007) Thermal activated (thermal) battery technology. Part II. Molten salt electrolytes. J Power Sources 164:397–414. https://doi.org/10.1016/j.jpowsour.2006.10.080

Study on Emulsion Phenomena and Field Flow Pattern in Side-Blown Copper Smelting Process

Xiao-long Li, Ting-an Zhang, Yan Liu and Dong-xing Wang

Abstract Side-blown process is regarded as an efficient, energy-saving, environment-friendly copper smelting technology. The arrangement of nozzles has significant impact on the distribution of emulsion droplets and flow pattern, which mainly decide the efficiency and result of the smelting process. In this paper, physical model was set up based on the same modified Froude number between model and industrial production. Effects of three kinds of nozzles on the size of emulsion droplet, field flow pattern and mixing time were analyzed. The results show: the SMD (Sauter mean diameter) of emulsion droplets turns normal distribution and the average SMD ranges from 2 to 4 mm. The size of emulsion droplets will decrease and the mixing time will be shorter when nozzles are more intensive. Two circulations exist in the fluid and the dead region (velocity below 0.05 m/s) will shrink when nozzles are arranged intensively.

Keywords Copper smelting process · Physical simulation · Emulsion droplets Flow pattern · Mixing time

Variables

A_i the interfacial area between two phases
d_0 Diameter of the injection nozzle (m)

X. Li (✉) · T. Zhang · Y. Liu · D. Wang
Key Laboratory of Ecological Utilization of Multi-Metal Intergrown Ores of Ministry of Education, School of Metallurgy, Northeastern University, Shenyang, China
e-mail: leedragong@163.com

T. Zhang
e-mail: zta2000@163.net

Y. Liu
e-mail: shanqibao2000@163.com

D. Wang
e-mail: 710797438@qq.com

G. Lambotte et al. (eds.), *Materials Processing Fundamentals 2018*, The Minerals, Metals & Materials Series,
https://doi.org/10.1007/978-3-319-72131-6_5

d_{32} the Sauter mean diameter of droplets
d_i the equivalent diameter of a single droplet
Fr_m Modified Froude number
g Gravitational acceleration (m s^{-2})
Q Gas flow rate (m^3 s^{-1})
V_d the volume of the emulsified droplets

Greek

ρ Density (kg m^{-3})

Introduction

Copper smelting furnaces can be divided into top, side and bottom blowing furnaces according to the position of oxygen lances. The oxidation rate of copper concentrate and the utilization of oxygen depend essentially on the interaction between slag and matte created by oxygen-enriched air stirring. To enhance the interaction among copper concentrate, slag and matte and promote the oxidation reaction, oxygen-enriched air is often injected into molten through the top, side or bottom lances of a smelting furnace. In this process, a gas phase whose motion is controlled by the initial force forms when gas enters the reactor at the beginning. The bubbles that the buoyancy dominates their movement form at the end of the trajectory. The rising bubbles create good hydrodynamic conditions, pushing the molten upward. They will also accelerate the oxidation reaction by stirring the slag-metal interface, which create the emulsification of slag into molten matte and significantly increase the contact area between slag and matte.

Both hydrodynamics and the associated transport phenomena of multiphase flows are of great significance for the production of high quality matte. Yan et al. [1] adopted Realizable k-ε model to numerically simulate the changes of hydrodynamic parameters in bottom-blowing bath smelting process. Yu et al. [2] investigated the effect of nozzle structure on bottom-blowing furnace flowing process. The percentage of stirring zone for four kinds of nozzle structure ranged from 17.07 to 18.95%, and the pressure fluctuation of A plane was larger, so the oxygen lance was more abrasive. Zhan et al. [3] took the interphase slip and buoyancy force into account and developed CFD method to describe the flow pattern in side-blown metallic bath. They found that although the side blown is only in one side, the major reactive region is still so large that can guarantee the quick metallurgical reactions. Bath can also absorb the strike from gas injection and weaken the splash. The big circulation flow and small vortexes create the good dynamic conditions for mass and heat transfers. Feng et al. [4] carried out the physical model experiments to clarify the effect of nozzle arrangements on mixing time in high efficient copper side-blown smelting reactor. The mixing time decreases when the nozzles are arranged intensively. While the mixing time will prolong when the diameter of nozzle becomes larger. Even though a number of papers [5–7] have been published to understand the effect of gas stirring, nozzle arrangement and other operating parameters on the flow patterns and fluid mixing performance in copper smelting process, few papers related to flow pattern in copper side-blown smelting process can be found.

The emulsification that contains sulphides, metal particles and smelting fume plays a crucial role in the oxidation reaction and the separation process between slag and matte. Lachmund et al. [8] proposed that emulsified slag droplets into the bulk of metal may occur when the gas flow rate exceeds a critical limit. The emulsified droplets accelerate the reaction rate due to their number density, size distribution and residence time in liquid metal. Misra et al. [9] developed an improved procedure to evaluate the viscosity of gas-slag foam at high gas void fractions and calculated velocity of gas bubbles in foaming slags. They developed a dynamic slag foaming model and applied it successfully to different converters when the nozzle designs and blowing regimes are clear. Khajavi et al. [10] pointed out that the rate of mass transfer between two phases is directly decided by the interfacial area, but few direct measurements of interfacial area has been achieved due to the experimental difficulties. They used a high speed CCD camera to study emulsification and interfacial area increase in bottom-blown metallurgical process by water modeling. The droplet size distribution and water-oil interfacial area increase were compared between two different water-oil systems.

Side-blown copper smelting process is an environmental-friendly and energy saving technology. Injecting oxygen-enriched air to the smelting furnace from two sides is the prominent operating feature. The air stirs the slag-matte interface intensely and accelerates the mixing process of copper concentrate into molten, which reduces the oxidation time and smoke emission. Meanwhile, the emulsified layer can separate slag from emulsion droplets by density difference between slag and copper matte or free mass diffusion created by concentration difference. The stable emulsified phenomena play an essential role in low copper content in slag, shortening the copper slag depletion process.

The purpose of the present study is to analyze the emulsification and flow patterns in side-blown copper smelting process. PIV (Particle Image Velocimetry), stimulus-response and static image collection techniques are utilized to investigate the effect of oxygen lance arrangement on emulsified droplets size distribution, fluid mixing performance and flow field in physical model experiment.

Theory

Dynamic Similarity

When gas is injected into liquid continuously, the two-phase area can be divided into three distinct regimes: the momentum, transition and buoyancy regimes [11, 12].

In momentum regime, the momentum transfer from the gas injection causes the liquid flow. The motion of the liquid in buoyancy regime is caused by the buoyancy of rising bubbles. Both momentum transfer and buoyancy predominate the liquid motion in transition regime. To guarantee the similar flow regimes in two systems, the modified Froude number that represents the ratio of the initial inertia force of injected gas to the buoyancy force need to be the same. Marco [13] investigated the hydrodynamics of air/water plumes in a large-scale model of a metallurgical ladle which was a cylindrical vessel 1600 mm in ID and 2250 mm in total height. They correlated the trajectory with by the modified Froude number which was obtained by non-dimensionalizing the momentum equation.

$$Fr_m = \frac{\rho_g}{\rho_l} \frac{Q^2}{g d_0^5} \tag{1}$$

Thus, the gas flow rate Q in model can be decided when the modified Froude numbers are the same in model and industrial production.

The Equal Interfacial Tension

Interfacial tension which equals the free energy per unit area of the interface between slag and matte phases, determines the size distribution of emulsified droplets after the gas stirring [14]. When two emulsified droplets collide, the interfacial tension also decides whether they coalesce or not. The interfacial tension of oil-water in model should be equal to the value of slag-matte in industry.

Sauter Diameter of Emulsified Droplets

Assuming that the droplets are spherical, the following equations can be adopted to calculate the Sauter mean diameter (d_{32}) and the interfacial area (A_i) when the size distribution of the dispersed droplets were measured [10].

$$d_{32} = \frac{\sum d_i^3}{\sum d_i^2} \tag{2}$$

$$A_i = \frac{6V_d}{d_{32}} \tag{3}$$

Experimental Works

Experimental System and Parameters

A water bath covered with a layer of oil (100 mm in height) to represent the slag was used for the experiments. The water (about 200 mm-height) was poured into the physical model of side-blown copper smelting furnace. Schematic of the whole experimental system is depicted in Fig. 1. The bath is stirred by the injecting compressed air through the nozzles located at two sides of the tank. The diameter of the nozzle is 3.75 mm. The properties of the fluid (water, oil and air in model; matte, slag and oxygen-enriched air in industry) and the operating parameters are shown in Table 1.

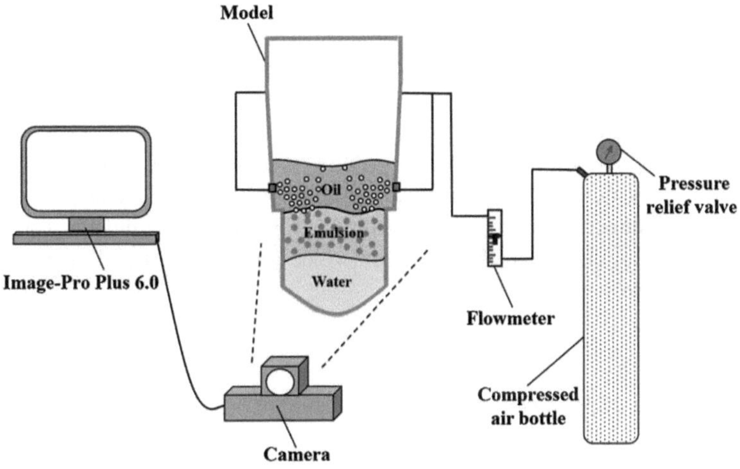

Fig. 1 Schematic diagram of whole experimental system

Table 1 Main physical parameters in smelting furnace and physical model

Fluid	Density (kg/m^3)	Viscosity (Pa s)	Temperature (K)	Flow rate (Nm^3/h)	Pressure (MPa)	Interfacial tension (N/m)	Flow rate (Nm^3/h)
Enriched oxygen	1.187	1.921×10^{-5}	283–323	8400–9400	0.09–0.1	–	8074–9285
Smelting slag	3220	1.25	1180–1250	–	–	0.02–0.06	–
Copper matte	4490	3.8×10^{-3}	1150–1180	–	–		–
Air	1.146	1.861×10^{-5}	298–303	17–23	0.1	–	20–23
Oil	890	5.4×10^{-5}	298–303	–	–	0.024	–
Water	998	8.937×10^{-4}	298–303	–	–		–

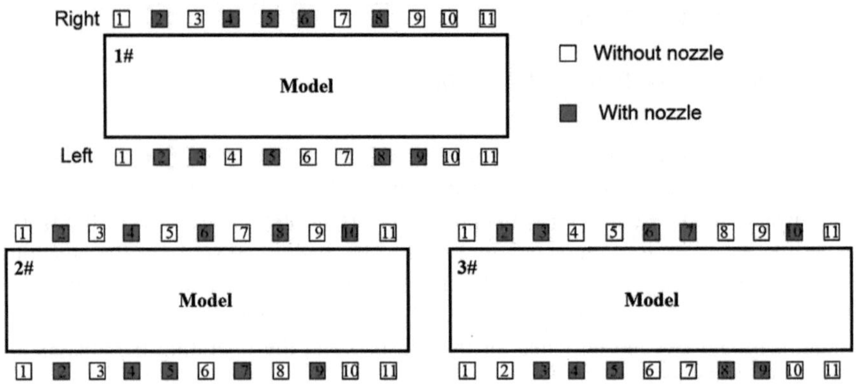

Fig. 2 Three kinds of nozzle arrangement

Figure 2 shows the different nozzle arrangements. From the overhead view of the model, each side has 11 lances. But only 10 lances in total (5 ones each side) will be used in production. The oxygen lance acts as the power source and the supplier of reactant, which directly determines the motion of liquid, the size distribution of the emulsion droplets and the trajectory of bubbles. This paper aims to investigate the effect of nozzle arrangement on flow field and emulsification, so the intensive lance arrangement 3#, the dispersive configuration 2# and the transitional one 1# are chose.

Principles of Measurements

The interfacial area is a function of the volume and size distribution of emulsion droplets. In order to quantify the interfacial area, a CCD camera was used to record the changes of the emulsified layer. The volume of the emulsion is calculated by detecting the height of emulsion, so the total volume of emulsified droplets (V_d) can be estimated when the gas injection stopped. The size distribution of droplet can be got by analyzing the still images after the steady state was reached. About 300 emulsion droplets were analyzed and the Sauter mean diameter can be calculated.

PIV (Particle Image Velocimetry) technique is utilized in the measurement of the liquid flow pattern. The laser device was located at one side of the model and shot two laser beams which have a constant offset time. The CCD camera was arranged in front of the model, which was vertical to the laser beam. When the flow was steady, 40 images were recorded during 4 s and the flow pattern was analyzed by the image processing mode in PIV device.

The mixing time was measured by the conductivity meter. Figure 3 shows the calculating method of the mixing time for fluid. When the fluid flow steadily, saturated KCl solution was poured into the model from the top charging hole. From

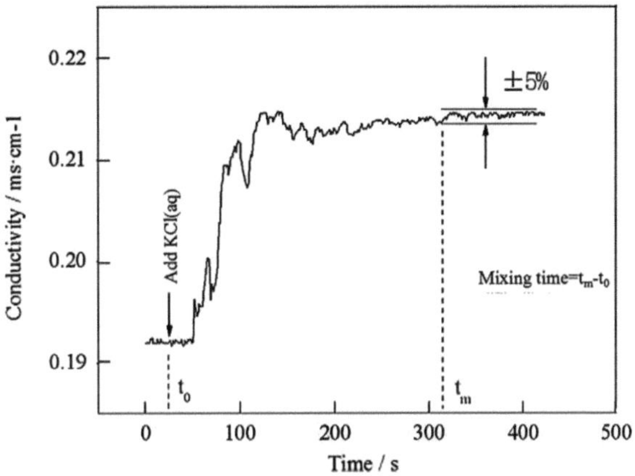

Fig. 3 The method of calculating the mixing time of fluid

this moment on, the change of fluid conductivity with time was recorded and the experiment was stopped when the conductivity fluctuated at a constant (the amplitude not exceeding 5%). The time interval, namely, is the mixing time.

Results and Discussion

Emulsified Droplet Size Distribution and Interfacial Area

Any two-phase reaction is considerably accelerated by gas stirring, as a result of increasing the interfacial area. Interfacial area of the emulsified region is calculated by substituting the value of total volume of droplets in emulsified region (V_d)—obtained from measuring the height of the emulsified region—back into Eq. 3. The equivalent diameter of the droplets against the nozzle arrangement is plotted in Fig. 4. As seen in Fig. 4, although the equivalent diameter ranges from 0.5 to 7.0 mm, the highest peak occurs around 3.0 mm. The difference between the relationship of d_i and nozzle arrangement in water-oil system is explained as follows. The size of a droplet is directly determined by the shear stress of the adjacent moving fluid, and the motion of the fluid is controlled by the gas injection from nozzles located at two sides of the model. For 3#, three or two nozzles are arranged together (seen from the 2–3, 6–7 in the right; 3–4–5, 8–9 in the left in Fig. 2), so it has the most intensive nozzle arrangement. In this case, the energy of gas is more concentrated, which creates the bigger inertia and magnitude of stress, breaking the oil phase into many smaller emulsified droplets. While for 2#, nozzles are arranged alternately, as the 2–4–6–8–10 at the right side, so the momentum of injected gas is dispersed, leading to the lower stress force. From the 2# in Fig. 4, the curve is lower than others and the obvious peaks at 4.0 and 5.5 mm can be seen. The lower magnitude of stress creates more droplets with less inertia, and the velocity at the interface moves the droplets around instead of breaking them. The cumulative

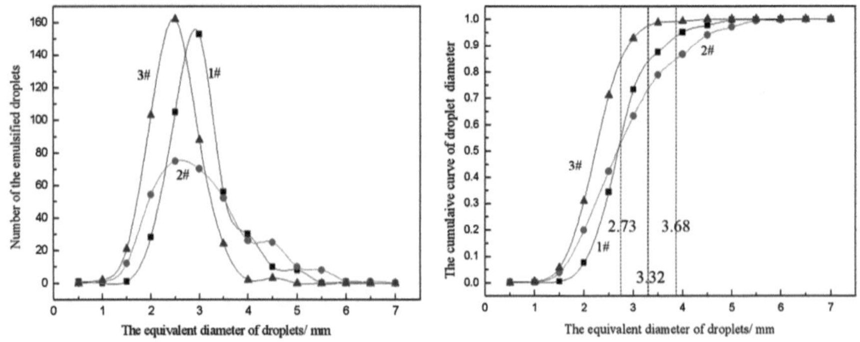

Fig. 4 The size distribution and the cumulative curve of emulsified droplets

Table 2 Sauter mean diameter and interfacial area under different nozzle arrangements

Items	d_{32} (mm)	V_d (m^3)	A_{il} (m^2)
1#	3.32	1.589×10^{-2}	28.72
2#	3.68	1.301×10^{-2}	21.22
3#	2.73	1.178×10^{-2}	25.88

curves of the equivalent of the emulsified droplets can also be seen in Fig. 4. No. 3 nozzle arrangement has the steepest curve ranging from 1.5 to 3.0 mm and about 90% of droplets are smaller than 3.0 mm. With the nozzle becoming dispersed, the slope of curves turn to smooth, and the size distribution becomes more extensive. 90% of droplets distribute in 0.5–3.75 mm and 0.5–4.2 mm for 1# and 2#, respectively.

Sauter mean diameter of the emulsified droplets not only determines the interfacial area in emulsified region, but also has a substantial influence in the residence time of the droplets. After the formation of a droplet in emulsion layer, it will move upwards. The buoyancy, gravity and viscosity resistance need to be considered in the vertical movement of the droplet. When the size of droplet exceeds a limit, the buoyancy controls the motion and will accelerate the floating process. So the residence time of larger droplets is shorter than that of the small ones, more large droplets will return to the upper slag phase. As a result, the volume of the emulsified region will shrink and the interfacial area will also decrease. Table 2 shows the Sauter mean diameter of the droplets in emulsified region and the volume of the emulsified region against the nozzle arrangement. 3# has the smallest value of 2.73 mm and the smallest volume of the emulsified region, so it has the intermediate interfacial area. While 2# has the biggest SMD value (up to 3.68 mm) and the intermediate value for emulsified region, so it has the slowest value for interfacial area. Although 1# has the intermediate SMD value which is lower than the figure of 3#, 1# has the biggest emulsified volume, leading to the biggest interfacial area for 1#.

Flow Pattern and Mixing Time

Figure 5 illustrates the different flow patterns of fluid against three kinds of nozzle arrangement. Two big circulations that occupy the lower fluid occur in all three cases. The fluid flows downward close to the side wall and then flows upwards, thus forming the circulations. In the case of 1# and 3#, two circulations develop fully, but the right one in 2# is so small that only takes up the 1/5 space and leas to the big dead area in the central region of the model. Compared to the 3#, two circulations in 1# flow quickly and turn more intensive. The dead area is smaller than that of 3#. The motions of the upper fluid are too complex and splash also exists in all three cases due to the violent agitation by gas injection. In the industrial production, the expected flow patterns may be the intense mixing performance in the upper liquid and the peaceful flow for lower liquid. The gas can stir the upper liquid strongly, which guarantees the perfect mass transfer and efficient oxidation reaction between slag and copper concentrate. Meanwhile, the intensive gas stirring can also break the emulsified droplets into more smaller ones, supplying the bigger interfacial area and longer residence time. The peaceful flow in lower fluid will not mix the slag and matte together again, ensuring the low copper content in slag and high concentration matte.

For a multiphase reaction system, the shorter mixing time means the longer reaction time when the residence time of the reactant is the same. The mixing time is an important parameter for efficient reactor design. The mixing time, mass transfer time and reaction time are three essential factors that evaluate the controlling mechanism of the process. For the side-blown copper smelting process, the efficient mixing is necessary. The shorter mixing time, the more efficient gas momentum transfer. The mixing time against different nozzle arrangements can be seen in Fig. 6. The values are over 400 s, and the shortest mixing time is about

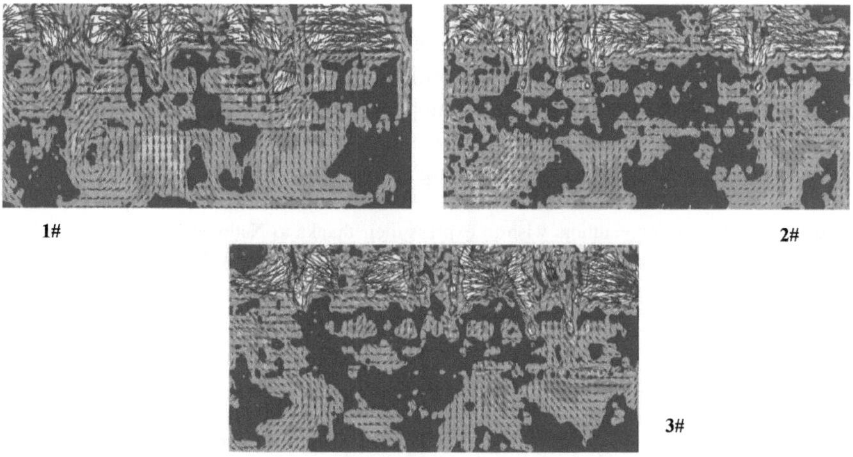

Fig. 5 The flow patterns of fluid against different nozzle arrangements

Fig. 6 The mixing time of fluid under different nozzle arrangements

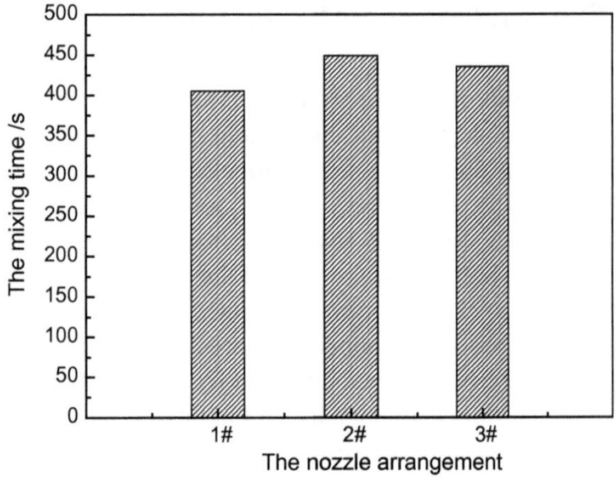

400 s in 1#. The longest mixing process occurs in 2# with the loose nozzle arrangement. The intensive nozzle arrangement will shrink the mixing time and accelerate the mass transfer for side-blown copper smelting process.

Conclusions

The objectives of this work were to study the emulsion phenomena and field flow pattern in side-blown copper smelting process. The CCD camera recorded the emulsified region under three kinds of nozzle arrangement. The equivalent diameters of emulsified droplet mainly distribute from 1.0 to 5.0 mm, and the smallest SMD value is 2.73 mm for the intensive nozzle arrangement-3#, the increasing figure 3.68 for the loose nozzle arrangement-2#. No. 1 nozzle arrangement has the largest emulsified volume, and its interfacial area can reach 28.72 m^2.

The fluid flow patterns and the mixing time were measured by PIV and stimulate-response techniques. Two circulations can be found in all three nozzle arrangements, and the motions of the upper fluid turn more complex. When the nozzles are arranged loosely, the inertia of gas becomes weak and the circulation will shrink, which leads to longer mixing time.

Acknowledgements The authors wish to express their thanks to National Natural Science Fund Committee, which provided financial support for this work (No. U1402271).

References

1. Yan H, Fangkan L, Zhenyang Z et al (2012) Influence of lance arrangement on bottom-blowing bath smelting process. Chin J Nonferrous Met 22:2393–2400

2. Yu Y, Zhi W, Xunliang L et al (2014) Simulation and experiment of influence of nozzle structure on bottom-blowing furnace flowing process. J Cent South Univ (Science and Technology) 45:4129–4137
3. Shuhua Z, Chaobin L, Tseqiang H (2003) CFD analysis of gas stirring behavior in side-blown metallic bath. J Cent South Univ Technol 34:148–151
4. Feng Y, Jun Z, Xiaolong L et al (2015) Physical simulation on mixing time in high efficient copper side-blown smelting reactor. Nonferrous Met (Extractive Metallurgy) 6:1–3, 21
5. Yamashita S, Iguchi M (2003) Control of reverse emulsification and mixing time in a bottom blown bath covered with top slag. ISIJ Int 43:1326–1332
6. Li B, Yin H, Zhou CQ et al (2008) Modeling of three-phase flows and behavior of slag/steel interface in an argon gas stirred ladle. ISIJ Int 48:1704–1711
7. Tang J, Wang N, Shyy W (2011) Flow structures of gaseous jets injected into water for underwater propulsion. Acta Mech Sin 27:461–472
8. Helmut L, Xie Y, Buhles T et al (2003) Slag emulsification during liquid steel desulphurisation by gas injection into the ladle. Steel Res 74:77–85
9. Misra P, Brahma DEO, Chhabra RP (1998) Dynamic model of slag foaming in oxygen steelmaking converters. ISIJ Int 38:1225–1232
10. Khaiavi LT, Barati M (2010) Cold model study of emulsification behavior in bottom blown metallurgical baths covered with thick slag. ISIJ Int 50:654–662
11. Iguchi M, Takeuchi H, Morita Z (1991) The flow field in air-water vertical bubbling jets in a cylindrical vessel. ISIJ Int 31:246–253
12. Iguchi M, Nozawa K, Morita Z (1991) Bubble characteristics in the momentum region of air-water vertical bubbling jet. ISIJ Int 31:952–959
13. Castello-Branco MASC, Klaus S (1994) Large-scale measurements of the physical characteristics of round vertical bubble plumes in liquids. Metallurgical and materials transactions B 25B:359–371
14. Tan Y, Deng X, Liu T et al (2017) Influence of NaCl on the oil/water interfacial and emulsifying properties of walnut protein-xanthan gum. Food Hydrocolloids 72:73–80

Study on Minimum Starting Energy of Self-stirring Reactor Driven By Pressure Energy

Zimu Zhang, Qiuyue Zhao, Maoyuan Li, Xuhuan Guo, Dianhua Zhang and Ting-an Zhang

Abstract Stirring plays an important role in the metallurgical process. Enhancing the agitation of materials can strengthen the turbulence in multiphase flow, which can shorten uniform mixing time and reaction time. In order to improve leaching reaction and heat transfer efficiency, a novel self-stirring tubular reactor driven by pressure energy used in bauxite digestion process is presented originally by Northeastern University, which combines the advantages of autoclaves and conventional tubular digestion equipment. According to the principle of dimensional analysis, the dimensionless equation between input power and various factors is established, and it is corrected to obtain input power, through verified, the calculated values are basically consistent with the experimental results.

Keywords Self-stirring · Dimensional analysis · Starting energy Reactor

Introduction

Stirring plays an important role in metallurgy, chemical and material processing units operation. Enhancing the agitation of materials can strengthen the turbulence in multiphase flow, which can shorten uniform mixing time and reaction time. The increase of turbulence intensity can also effectively control the adhesion of scar on

Z. Zhang (✉) · Q. Zhao · M. Li · X. Guo · T. Zhang
School of Metallurgy & Key Laboratory of Ecological Utilization of Multi-metal Intergrown Ores of Ministry of Education, Northeastern University, Shenyang 110819, China
e-mail: zhangzimu111@sina.cn

Q. Zhao
e-mail: zhaoqy@smm.neu.edu.cn

T. Zhang
e-mail: zta2000@163.com

Z. Zhang · D. Zhang
State Key Lab of Rolling & Automation, Northeastern University, Shenyang 110819, China

G. Lambotte et al. (eds.), *Materials Processing Fundamentals 2018*,
The Minerals, Metals & Materials Series,
https://doi.org/10.1007/978-3-319-72131-6_6

heat-transfer surface. For high temperature process, agitation also enhances heat and mass transfer [1, 2]. However, the biggest problem of the traditional mechanical stirred tubular reactor is that it requires higher mechanical seals and is not conducive to a substantial increase in temperature and pressure. Therefore, how to strengthen the heat and mass transfer of the reactor and reduce scarring without increasing energy consumption is the development direction of the pressurized leaching reactor [3–7].

To solve the above problems, based on years of research the Northeastern University proposes a self-stirring reactor dissolution type [8], which can drive the stirrer rotation through a high pressure fluid self energy and realize the mixing effect on reaction medium, strengthening the reaction process. Method of dimensional analysis is adopted in this paper to study the relationship between start energy of self-stirring reactor and various factors, and to analyze specifically the relationship between input power (input energy per unit time) and various factors.

Determination of Formula of Quasi Numerical Equations

Input power N is mainly affected by the following factors: liquid density ρ, liquid viscosity μ, impeller diameter D, reactor diameter d, blade thickness δ, blade number n, inlet pressure P, inlet velocity u, liquid level H and stirring speed R. Then a general functional relation can be listed as:

$$N = f(\rho, \mu, D, d, \delta, n, P, u, H, R) \tag{1}$$

The general form of dimensional expression is:

$$[x] = L^l T^t M^m \tag{2}$$

In the formula, $[x]$ represents dimension of physical quantity X, L is dimension of length, T is dimension of time and M is dimension of mass. The dimensional expressions of all physical quantities can be represented by Formula (2). Dimension of variables in Formula (1) is shown in Table 1.

Based on analysis principle of theorem, $n - k = 8$ dimensionless combinations can be established. P, D and μ are selected as independent variables. Each Π is represented as

Table 1 Table of variable dimension

	N	ρ	μ	D	d	δ	n	P	u	H	R
M	1	1	1	0	0	0	0	1	0	0	0
L	2	−3	−1	1	1	1	0	−1	1	1	0
T	−3	0	−1	0	0	0	0	−2	−1	0	−1

$$\Pi_0 = \mu^{\alpha_0} D^{\beta_0} P^{\gamma_0} N \quad (3)$$

$$\Pi_3 = \mu^{\alpha_3} D^{\beta_3} P^{\gamma_3} \mathrm{d} \quad (4)$$

$$\Pi_4 = \mu^{\alpha_4} D^{\beta_4} P^{\gamma_4} L \quad (5)$$

$$\Pi_5 = \mu^{\alpha_5} D^{\beta_5} P^{\gamma_5} \mathrm{n} \quad (6)$$

$$\Pi_7 = \mu^{\alpha_7} D^{\beta_7} P^{\gamma_7} H \quad (7)$$

$$\Pi_8 = \mu^{\alpha_8} D^{\beta_8} P^{\gamma_8} R \quad (8)$$

$$\Pi_9 = \mu^{\alpha_9} D^{\beta_9} P^{\gamma_9} \rho \quad (9)$$

$$\Pi_{10} = \mu^{\alpha_{10}} D^{\beta_{10}} P^{\gamma_{10}} \mathrm{u} \quad (10)$$

For Π_0, dimensions of μ, D, P and N are plugged into Formula (3) to obtain:

$$M^0L^0T^0 = \left[ML^{-1}T^{-1}\right]^{\alpha_0}[L]^{\beta_0}\left[ML^{-1}T^{-2}\right]^{\gamma_0} ML^2T^{-3} \quad (11)$$

Thus exponential equation can be obtained:

$$\begin{aligned} M &: 0 = \alpha_0 + \gamma_0 + 1 \\ L &: 0 = -\alpha_0 + \beta_0 - \gamma_0 + 2 \\ T &: 0 = -\alpha_0 - 2\gamma_0 - 3 \end{aligned} \quad (12)$$

Solutions are $\alpha_0 = 1$, $\beta_0 = -3$, $\gamma_0 = -2$, therefore, $\Pi_0 = \mu D^{-3} P^{-2} N$

Similarly, for Π_3, dimensions of μ, D, P, d are plugged into Formula (4) to obtain:

$$M^0L^0T^0 = \left[ML^{-1}T^{-1}\right]^{\alpha_3}[L]^{\beta_3}\left[ML^{-1}T^{-2}\right]^{\gamma_3} L \quad (13)$$

Thus exponential equations can be obtained:

$$\begin{aligned} M &: 0 = \alpha_3 + \gamma_3 \\ L &: 0 = -\alpha_3 + \beta_3 - \gamma_3 + 1 \\ T &: 0 = -\alpha_3 - 2\gamma_3 \end{aligned} \quad (14)$$

Solutions are $\alpha_3 = 0$, $\beta_3 = 1$, $\gamma_3 = 0$, so $\Pi_3 = D^{-1}d$

Similarly, for Π_4, dimension of μ, D, P, L are put into Formula (5) to obtain:

$$M^0L^0T^0 = \left[ML^{-1}T^{-1}\right]^{\alpha_4}[L]^{\beta_4}\left[ML^{-1}T^{-2}\right]^{\gamma_4}L \tag{15}$$

So exponential equations can be obtained:

$$\begin{aligned} M &: 0 = \alpha_4 + \gamma_4 \\ L &: 0 = -\alpha_4 + \beta_4 - \gamma_4 + 1 \\ T &: 0 = -\alpha_4 - 2\gamma_4 \end{aligned} \tag{16}$$

Solutions are $\alpha_4 = 0$, $\beta_4 = 1$, $\gamma_4 = 0$, so $\Pi_4 = D^{-1}L$
Similarly, for Π_5, dimension of μ, D, P, n are put into Formula (6) to obtain:

$$M^0L^0T^0 = \left[ML^{-1}T^{-1}\right]^{\alpha_5}[L]^{\beta_5}\left[ML^{-1}T^{-2}\right]^{\gamma_5} \tag{17}$$

So exponential equations can be obtained:

$$\begin{aligned} M &: 0 = \alpha_5 + \gamma_5 \\ L &: 0 = -\alpha_5 + \beta_5 - \gamma_5 \\ T &: 0 = -\alpha_5 - 2\gamma_5 \end{aligned} \tag{18}$$

Solutions are $\alpha_5 = 0$, $\beta_5 = 0$, $\gamma_5 = 0$, so $\Pi_5 = n$
For Π_7, dimension of μ, D, P, $H$are plugged into Formula (7) to obtain:

$$M^0L^0T^0 = \left[ML^{-1}T^{-1}\right]^{\alpha_7}[L]^{\beta_7}\left[ML^{-1}T^{-2}\right]^{\gamma_7}L \tag{19}$$

So exponential equations can be obtained:

$$\begin{aligned} M &: 0 = \alpha_7 + \gamma_7 \\ L &: 0 = -\alpha_7 + \beta_7 - \gamma_7 + 1 \\ T &: 0 = -\alpha_7 - 2\gamma_7 \end{aligned} \tag{20}$$

Solutions are $\alpha_7 = 0$, $\beta_7 = 1$, $\gamma_7 = 0$, so $\Pi_7 = HD^{-1}$
For Π_8, dimension of μ, D, P, R are put into Formula (8) to get:

$$M^0L^0T^0 = \left[ML^{-1}T^{-1}\right]^{\alpha_8}[L]^{\beta_8}\left[ML^{-1}T^{-2}\right]^{\gamma_8}T^{-1} \tag{21}$$

Thus exponential equations can be obtained:

$$\begin{aligned} M &: 0 = \alpha_8 + \gamma_8 \\ L &: 0 = -\alpha_8 + \beta_8 - \gamma_8 \\ T &: 0 = -\alpha_8 - 2\gamma_8 - 1 \end{aligned} \tag{22}$$

Solutions are $\alpha_8 = 1$, $\beta_8 = 0$, $\gamma_8 = -1$, so $\Pi_8 = \mu P^{-1}R$
For Π_9, dimension of μ, D, P, ρ are put into Formula (9) to obtain:

$$M^0L^0T^0 = \left[ML^{-1}T^{-1}\right]^{\alpha_9}[L]^{\beta_9}\left[ML^{-1}T^{-2}\right]^{\gamma_9}ML^{-3} \tag{23}$$

So exponential equations can be obtained:

$$\begin{aligned} M &: 0 = \alpha_9 + \gamma_9 + 1 \\ L &: 0 = -\alpha_9 + \beta_9 - \gamma_9 - 3 \\ T &: 0 = -\alpha_9 - 2\gamma_9 \end{aligned} \tag{24}$$

Solutions are $\alpha_9 = -2$, $\beta_9 = 2$, $\gamma_9 = 1$, so $\Pi_9 = \mu^{-2}D^2P\rho$
For Π_{10}, dimension of μ, D, P, u are plugged into Formula (10) to obtain:

$$M^0L^0T^0 = \left[ML^{-1}T^{-1}\right]^{\alpha_{10}}[L]^{\beta_{10}}\left[ML^{-1}T^{-2}\right]^{\gamma_{10}}LT^{-1} \tag{25}$$

So exponential equations can be got:

$$\begin{aligned} M &: 0 = \alpha_9 + \gamma_9 \\ L &: 0 = -\alpha_9 + \beta_9 - \gamma_9 + 1 \\ T &: 0 = -\alpha_9 - 2\gamma_9 - 1 \end{aligned} \tag{26}$$

Solutions are $\alpha_{10} = 1$, $\beta_{10} = -1$, $\gamma_{10} = -1$, so $\Pi_{10} = \mu D^{-1}P^{-1}u$
Thus the following formula can be got:

$$f\left(\mu D^{-3}PN, \mathrm{d}D^{-1}, LD^{-1}, n, HD^{-1}, \mu P^{-1}R, \mu^{-2}D^2P\rho, \mu D^{-1}P^{-1}u\right) = 0 \tag{27}$$

The upper formula can also be expressed as

$$f_1 = \left(\mu D^{-3}PN, \mathrm{d}D^{-1}, LD^{-1}, n, HD^{-1}, \mu P^{-1}R, \mu^{-2}D^2P\rho, \mu D^{-1}P^{-1}u\right) \tag{28}$$

To obtain the expression for the input power N, Formula (4–28) is expressed as an explicit function form:

$$\mu D^{-3}PN = f_2\left(\mathrm{d}D^{-1}, LD^{-1}, n, HD^{-1}, \mu P^{-1}R, \mu^{-2}D^2P\rho, \mu D^{-1}P^{-1}u\right) \tag{29}$$

According to the specific experimental conditions of this experiment, the diameter of impeller blade D, blade thickness L, blade number n and mixer diameter d are fixed values. So the following formula can also be got:

$$\mu D^{-3}PN = f_2\left(HD^{-1}, \mu P^{-1}R, \mu^{-2}D^2P\rho, \mu D^{-1}P^{-1}u\right) \tag{30}$$

Formula (30) shows that there is a total of 8 variables in this formula: μ, D, P, N, H, R, u and ρ. According to the characteristics of equipment in the experiment, the

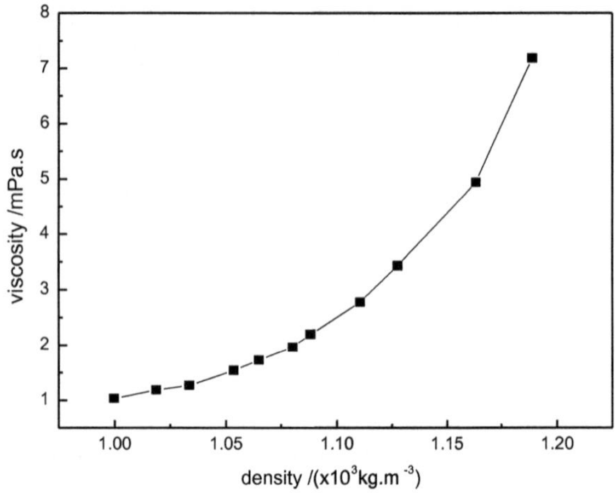

Fig. 1 Relations between density and viscosity

pressure P is in one-to-one correspondence with the inlet velocity u, so $u = f_P(P)$, which means flow rate is a function of pressure, so P and u can not be used as two independent variables, the dimensionless variable Π_{10} need to be taken away.

Similarly, the relationship between density and viscosity of fluid is measured, as shown in Fig. 1.

It is clear that viscosity and density of the fluid used in this paper are one-to-one correspondence, which means $\rho = f_\rho(\mu)$, so density ρ and viscosity μ also cannot exist as two independent variables, the dimensionless variable Π_9 also need to be taken away.

So Formula (30) is rewritten as:

$$\mu D^{-3}P^{-2}N = f_2\left(HD^{-1}, \mu P^{-1}R\right) \tag{31}$$

Determination of Relation of Quasi Number Equations

There are 6 variables in Formula (31): μ, D, P, N, H and R, where diameter D is fixed value, pressure P, liquid level H, viscosity μ and stirring speed R are measurable, N is input power. According to method of dimensional analysis, $N = PQ$ and P is pressure, the following equations are got:

$$\begin{cases} [N] = ML^2T^{-3} \\ [P] = ML^{-1}T^{-2} \\ [Q] = M^xL^yT^z \end{cases} \tag{32}$$

Solution is: $\begin{cases} x = 0 \\ y = 3 \\ z = -1 \end{cases}$

So$[Q] = L^3T^{-1}$, Q is the volume flow of fluid.

From measured experimental data the volume flow can be obtained:

$$Q = \frac{1}{5}\sum_{i=1}^{5}\frac{10^{-3}\,\mathrm{m}^3}{T_i} \quad (33)$$

Therefore, input energy of fluid is the sum of static pressure energy P and dynamic pressure energy $\frac{1}{2}\rho v^2$, and the formula of input power N can be obtained:

$$N = (P + \frac{1}{2}\rho v^2) \cdot Q \quad (34)$$

From the above analysis, the empirical criterion formula can be fitted:

$$\mu D^{-3}PN = \mathrm{K}(HD^{-1})^{\mathrm{X}_1}(\mu P^{-1}R)^{\mathrm{X}_2} \quad (35)$$

where K, X_1, X_2 are fitting coefficients.

Logarithm of both sides of the upper form is obtained:

$$\ln(\mu D^{-3}PN) = \ln\mathrm{K} + \mathrm{X}_1\ln(HD^{-1}) + \mathrm{X}_2\ln(\mu P^{-1}R) \quad (36)$$

According to the above relation formula, the fitting coefficients can be obtained by processing experimental data,

which are: K = 232.986, $X_1 = -0.42345$, $X_2 = 1.45931$.

The empirical formula is obtained by putting the fitting coefficients into Formula (32):

$$N = 232.986\mu^{0.45931}D^{3.42345}P^{0.54069}H^{-0.42345}R^{1.45931} \quad (37)$$

where N-input power, W; μ-liquid viscosity, Pa s; P-inlet pressure, Pa; H-liquid level, m; D-impeller diameter, m; R-stirring speed, r/s.

The relation obtained by the above method is got on basis of previous experiments [9]. Therefore, the relationship between input power and stirring speed and other influence factors studied by this experiment can be predicted by using the above relation with given physical parameters, main structure parameters and operating parameters. However, it should be pointed out that the use of dimensional analysis formula should be limited to the experimental conditions of this paper.

Verification of Dimensional Analysis Formula

The input power obtained from dimensional analysis is verified under the conditions of this experiment. By comparing the actual input power N_0 at different pressures and the N_1 calculated in accordance with Formula (37), it can be seen that there is a difference between actual power and theoretical power, which is due to great resistance loss in process of experiment.

It can be seen from Fig. 2 that fitted power is less than actual power and there is a certain difference ΔN between them. The reason may be that the resistance loss of actual equipment is great. Therefore, the results are corrected according to validation test data and the influence of new resistance loss is eliminated. The corrected formula is obtained:

$$N = 232.986\mu^{0.45931}D^{3.42345}P^{0.54069}H^{-0.42345}R^{1.45931} + 28.43965 \quad (38)$$

It can be seen from Fig. 3 that revised fitted formula agrees well with actual measured values.

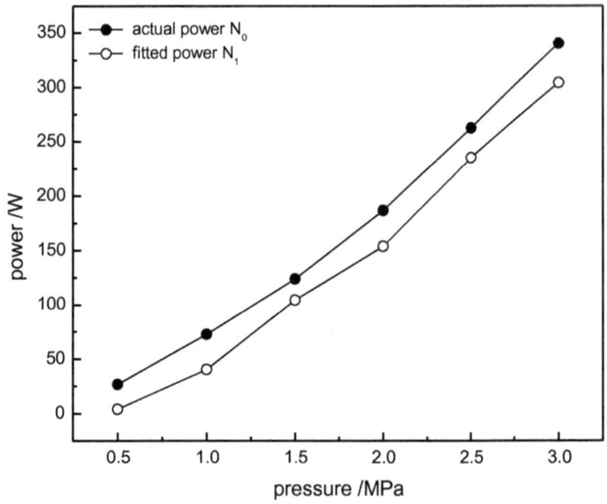

Fig. 2 Verification of dimensional analysis formula

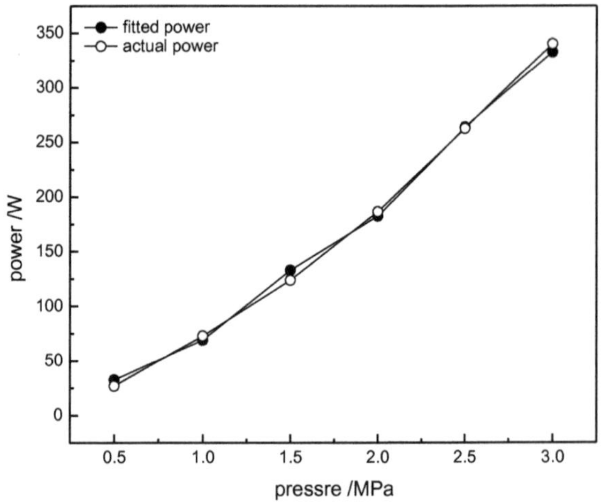

Fig. 3 Comparison of corrected fitted power and actual power

Conclusion

Method of dimensional analysis is adopted in this paper to study the relationship between start energy of self-stirring reactor and various factors. The main conclusions are as follows:

(1) According to the principle of dimensional analysis, the relationship between the input minimum energy of a self-stirring reactor with a certain rotational speed and various factors was deduced. The dimensionless equation between input energy of unit time, which is input power, and various factors is established: $\mu D^{-3}PN = \mathrm{K}(HD^{-1})^{X_1}(\mu P^{-1}R)^{X_2}$.

In the range of experimental research, the equation is solved and corrected, the input power is obtained, which is $N = 232.986\mu^{0.46}D^{3.42}P^{0.54}H^{-0.42}R^{1.46} + 28.44$, through verified, the calculated values are basically consistent with the experimental results.

Acknowledgements This research was supported by the National Natural Science Foundation of China (51204040, U1202274), Fundamental Research Funds for the Central Universities (N130607001), the National 863 Plan (2010AA03A405, 2012AA062303), the National Science and Technology Support Program (2012BAE01B02), National Program on Key Basic Research Project of China (973 Program) (2012CBA01205).

References

1. Huang K (2004) Research on extraction of platinum group metals by pressure cyanidation. Kunming University of Science and Technology, Kunming
2. Wang JK (2010) Application and development of pressure hydrometallurgy on zinc smelting. Non-ferrous industrial low carbon development–the national nonferrous metals industry conference on smelting low carbon economy and exhaust emissions, pp 186–188
3. Qiu DP (1994) Pressure hydrometallurgy process chemistry and industrial time. Mining Metall 3(4):55–67
4. Song FL, Ning MG (2001) The past, present and future of pressurized Hydrometallurgy. Hydrometallurgy 20(3):165–166
5. Lu XY (2008) Development status of metallurgical equipment industry in China in 2008. Shandong Metall 30(6):84–85
6. Zhai YC, Liu XH, Xu JZ (2001) Modern metallurgy. Electronics Industry Press, pp 55–56
7. Niu YJ (2002) Understanding of bauxite resources and development of alumina industry in China. Acad R:11
8. Zhang TA, Zhao QY, Zhang ZM et al (2013) A Kind of tubular self-stirring leaching reactor. China Patent
9. Zhang ZM, Zhao QY, Zhang DH et al (2015) A novel self-stirring tubular reactor used in bauxite digestion process. In: Light metals 2015. TMS, Warrendale, pp 27–34

Part IV
Alloy Processing and Properties Modeling

Yield Strength Prediction in 3D During Local Heat Treatment of Structural A356 Alloy Components in Combination with Thermal-Stress Analysis

Tobias Holzmann, Andreas Ludwig and Peter Raninger

Abstract The numerical investigation of common heat treated A356 as-cast parts was successfully exhibited by several authors for T4, T6 and T7 tempers. However, for non-uniform or local heat treatments (e.g. welding) the common procedure of estimating the material properties cannot be used. Therefore, a new methodology was developed in order to overcome the weak point in the kinetic calculation and to enable the prediction of the material properties in 3D. The single steps of the methods are described while mentioning interesting aspects briefly. Subsequently, a local heat treated structural A356 component case is presented while taking care about remelting and yielding. At the end possible areas of applications are given in which the new method can be applied.

Keywords Local heat treatment · Thermal-stress analysis · Yield strength prediction in 3D · Aluminum alloy

Introduction

Aluminum alloys are widely used in different industry sectors such as the automotive or aerospace ones. This is related to the excellent cast ability, high thermal and electric conductivity and corrosion resistance of aluminum alloys such as the A356 (AlSi7Mg0.3) alloy [1, 2]. The characteristic that these alloys can further be age hardened, which is mainly based on forming precipitates—typically magnesium included phases such as β or Θ and similar ones as the metastable β''—, during

T. Holzmann (✉) · A. Ludwig
Department of Metallurgy, Montanuniversitaet Leoben, Franz-Josef-Strae 18, A-8700 Leoben, Austria
e-mail: Tobias.Holzmann@unileoben.ac.at

P. Raninger
Material Center Leoben Forschung GmbH, Roseggerstrae 12, A-8700 Leoben, Austria

G. Lambotte et al. (eds.), *Materials Processing Fundamentals 2018*,
The Minerals, Metals & Materials Series,
https://doi.org/10.1007/978-3-319-72131-6_7

natural or artificial aging and thus be controlled in terms of material properties, support these alloys even more while offering an excellent strength to weight ratio [3]. In the past numerous experimental investigations were made to analyze the influence of different parameters onto the microstructure and mechanical properties of aluminum alloy. For example Wang et al. [4] analyzed the dependency of the pouring temperature and cooling condition during solidification on the as-cast microstructure and the behavior during remelting the eutectic regime. Prasada Rao et al. [5] used strontium and antimony modifier in combination with the Al5Ti2C grain modifier in order to manipulate the eutectic structure for short and long melt treatment time. This particular topic of modifying the eutectic using different modifier elements to influence the nucleation and therefore the microstructure was discovered by various authors [6–8]. Other scientists such as Salleh et al. [9] and Caceres et al. [10] analyzed the influence of the amount of different elements—e.g. magnesium—on the microstructure and mechanical properties of aluminum alloys. Borodianskiy et al. [3] proposed a new method to refine the eutectic regime using ceramic nanoparticles in combination of a gas-dynamic treatment. In addition, the precipitation sequence during aging is well known based on investigations of several authors such as Dutta et al. [11] or Edwards et al. [12] and is nowadays almost completely understood. Despite experimental investigations, numerical models were developed simultaneously to capture the kinetic phenomena during annealing and aging in order to estimate the material properties numerically. The models discover a wide range of fields e.g. nucleation, grain growth, precipitation hardening etc. and are published elsewhere [13–15]. Falahati et al. [17] compared numerical simulations and measurements for arbitrary heat treatment—natural and artificial aging of aluminum. The work of Wu et al. [15] compared a set of experimental data of an A356 alloy with numerical simulations too. Both were found in good agreement.

However, the numerical investigations are generally done using the common T4, T6 or T7 heat treatment sequences. Especially the analyze of the T6 pattern is widely presented. Referring to local heat treatments—welding included—the common numerical procedure to evaluate the material properties cannot be used.

One possible approach was already suggested by Ludwig and Holzmann [18]. However, this method does only give a rough estimation of the material properties and cannot be applied for complex geometries based on the high non-linear yield strength evolution during local heat treatments. Furthermore, the applied area-weighted interpolation method can only be used for simple 2D geometries. Therefore, a new framework is presented here allowing to calculate the material properties in 3D for either an approximation or a highly accurate solution. Additionally, there is no limitation in the choice of the alloy one wants to investigate or the application field, e.g. a short local heat treatment with or without subsequent quenching and aging or analyzing the material behavior during welding around the welding seam. The presented method also includes thermal-stress calculation in order to avoid yielding or remelting of the material during a local heat treatment.

General Model Description

For analyzing the influence of a local heat treated structural aluminum alloy onto the material properties, a new simulation framework has been developed in order to simulate the energy insertion and its diffusion into a structural part as well as the resulting thermal stresses. Both quantities are used to limit the energy insertion based on user defined functions prohibiting yielding and remelting of the structural as-cast part. This is done by evaluating permanently the maximum occurring Von Mises equivalent stress and temperature—in the structural part—against the basic yield strength and the solidus of the alloy. Reducing the energy insertion is done after a trigger value is exceeded either for the maximum Von Mises stress or the maximum temperature. During the heat treatment and to evaluate the material properties in 3D, temperature curves at random positions are tracked, stored and used in the kinetic calculations afterwards. Thus, the material properties of each tracked position is available and can be used to get the 3D yield strength distribution in the structural part. Figure 1 shows an overview of the model with an additional optimization procedure.

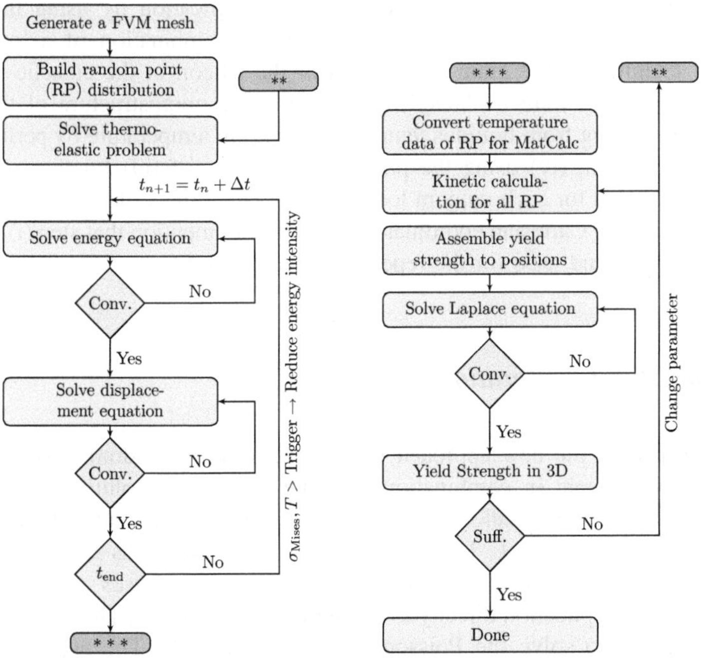

Fig. 1 General sequence of the new developed method for investigating into local heat treatments in combination with thermal-stress calculation and yield strength prediction in 3D. " Conv." stands for converged; " Suff." stands for sufficient

Energy and Stress Equations

The calculation of the energy distribution and the resulting thermal stresses during the local heat treatment are calculated using the finite volume method (FVM) approach. The general equations for the energy and displacements (for the stresses) are derived and implemented in the open source toolbox OpenFOAM® version 5 [19]. For reference, the equations are reported elsewhere [20–22]. The energy insertion is modeled as a boundary condition by using the multi-dimension Gaussian probability distribution [23]. Additional features of the boundary condition such as modeling convective heat transfer, moving heat sources (welding) etc. are explained in [24]. It is worth to mention that the displacement equation is calculated in a segregated manner without time derivative based on assumptions given in [20] and is not coupled with the energy equation due to small strains.

Kinetic Calculation

The evaluation of the material properties is done by using the scientific toolbox MatCalc [13, 16]. For the correct modeling approach, the numerical treatment is as follows: (i) estimate the phase fractions after solidification by using the Scheil approach; (ii) using the silicon phase fraction in combination of a log-normal function to initialize the grain distribution of the silicon in the eutectic—the log normal parameters were evaluated by analyzing the microstructure of the alloy; (iii) further cooling from liquidus temperature to room temperature by performing a kinetic calculation; (iv) using the phase compositions of (iii) for the subsequent kinetic evaluation for each random location; (v) extract the yield strength and store the result with the Cartesian coordinates. It is worth to mention that step (i) to (iii) is done only once and used for the repeated steps (iv) and (v).

Random Point Generation

The main part in the new approach is the generation of random positions in the structural component in combination with the subsequent solution of a Poisson equation. The random locations are used to track and store the temperature profiles during the heat treatment. Therefore, a new application was developed in the OpenFOAM® environment in order to use all in house classes and combine the outcome of the application directly to the thermal-stress solver. This also enables an easy procedure to solve the Poisson equation afterwards. The application itself determines the bounding box of the geometry in use and generates arbitrary locations within the numerical mesh automatically.

Two main strategies are developed right now. The user can either specify if the whole geometry or an own region is considered during the point generation. In addition, both variants can be combined while different point densities can be specified. This feature comes in handy if a more accurate solution of the yield strength is required locally.

The application itself generates a file containing cell IDs and the corresponding Cartesian positions of all points which are used in the local heat treatment step, and for manipulating the Poisson equation later on.

It should be noted that the application has the possibility to set a true random mode or not. Running the application in true random mode will produce a non-reproducible random point distribution while the alternative mode outputs a random distribution that can be re-generated by executing the application with the same settings again.

Yield Strength Distribution in 3D

Knowing the yield strength quantity at all random locations in combination with the position information allows one to calculate the yield strength distribution in 3D by solving a Poisson equation. The equation is solved for steady-state condition while the diffusion coefficient does not have a physical meaning and can be set arbitrarily —it influences only the amount of re-calculations. The source term on the right hand side of the equation ($\mathbf{Ax} = \mathbf{b}$) contains the known yield strength values at the corresponding positions. The numerical approach is implemented in OpenFOAM®.

It is worth to mention that instead solving the Poisson equation, FOAM offers the manipulation of the matrix system while adding explicit sources. Therefore, the Poisson equation can be replaced by the Laplace one. Based on the structure of the applications which were developed, the latter method is used.

Results and Discussion

An example of a local heat treatment is given now. A structural design, partly depicted in Fig. 2, is local heat treated around the bore hole by using the boundary condition mentioned previously. Before the thermal-stress calculation is performed, a random point distribution is generated. A possible outcome is shown in Fig. 4. Those points are used to track and store the individual temperature profiles.

During the simulation, the inserted energy is controlled in order to prevent yielding and remelting of the material; c.f. Figure 3. At the early stage (0–5 s) the energy insertion is controlled by using a parabola function in order to limit the maximal appearing Von Mises stress. The influence of the function is clearly visible in Fig. 3; the inserted energy profile follows the parabola function from 0 to 5 s. As the intensity of the LASER increases, the thermal stresses gain as well. After a

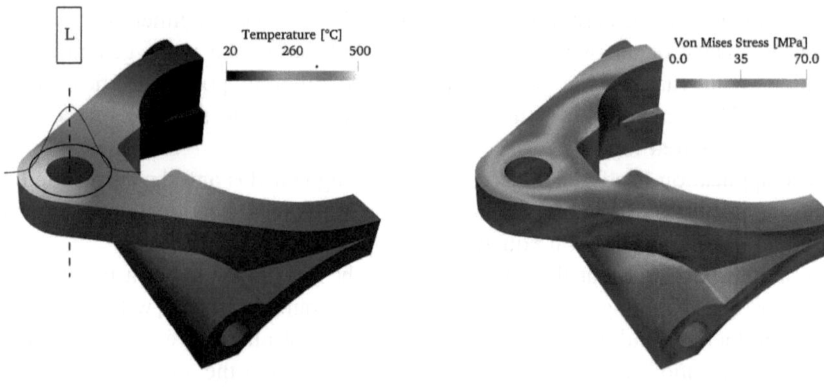

Fig. 2 Local heat treated structural part after 10 s. Left: Temperature profile and energy insertion; Right: Von Mises stress profile

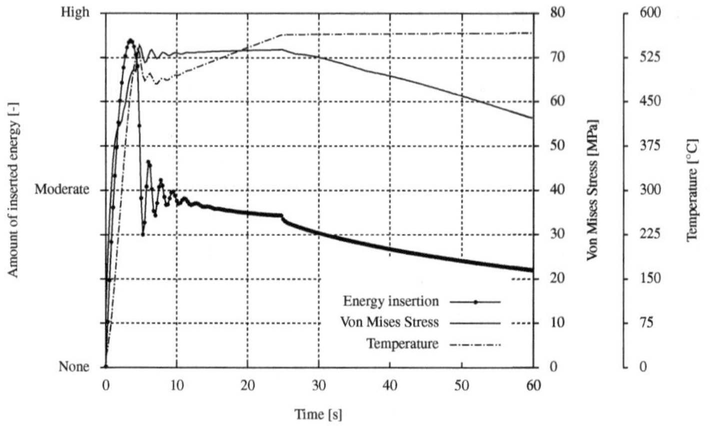

Fig. 3 Energy insertion control by the maximum occurring Von Mises stresses and temperature in the structural as-cast part. The energy control is based on user defined functions and are used after a trigger values is exceeded

defined Von Mises trigger value is reached the energy is reduced (5–25 s) by thermal stresses. Due to the high thermal conductivity—of the aluminum alloy—the temperature gradients get smoothed out rapidly and decrease the stresses respectively. In the meanwhile the temperature rises up to the temperature trigger value. Subsequently, the energy insertion is controlled by the temperature (25–60 s). A snap-shot of the temperature and stress profile at 10 s is depicted in Fig. 2 as well as a sketch of the laser direction. It is worth to mention that due to the reduced energy insertion, the highest temperature values are at the tail of the bore hole. This can be related to the high thermal conductivity and the accumulation of energy in that region. The Gaussian profile around the bore hole can only be observed in the first few seconds and is not shown here.

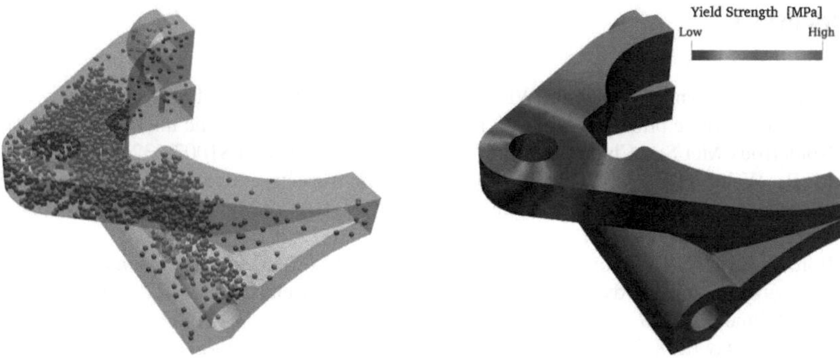

Fig. 4 Random point generation and yield strength distribution in 3D after the local heat treatment and artificial aging

After 60 s the whole part is water quenched (not shown in Fig. 3). The tracked temperature profiles of the random points are converted to a MatCalc readable file format and used for the kinetic calculations. After knowing the yield strength of the random points, the Poisson equation is solved and the predicted yield strength of the structural part is available in 3D cf. Figure 4.

Summary

The suggested method to estimate the yield strength distribution in 3D for structural aluminum parts can be used to investigate into local or non-uniform heat treatments while avoiding yielding and remelting. The intelligent connection between different software packages in combination with the new developed methodology allows further to investigate into different kinds of application field and alloys such as common heat treatment, local heat treatment and welding—investigating the behavior of the material around the welding seam.

Additionally, the 3D yield strength distribution can be used to optimize different areas of interest. E.g. one can calculate the volume averaged yield strength around the bore hole and using this quantity as an object function for an optimizing chain, c.f. Figure 1.

Acknowledgements Financial support by the Austrian Federal Government (in particular from Bundesministerium für Verkehr, Innovation und Technologie and Bundesministerium für Wirtschaft, Familie und Jugend) represented by Österreichische Forschungsförderungsgesellschaft mbH and the Styrian and the Tyrolean Provincial Government, represented by Steirische Wirtschaftsförderungsgesellschaft mbH and Standortagentur Tirol, within the framework of the COMET Funding Programme is gratefully acknowledged

References

1. Möller H, Govender G, Stumpf WE (2010) Application of shortened heat treatment cycles in A356 automotive brake calipers with respective globular and dendritic microstructures. Trans Nonferrous Met Soc China 20:814–820. https://doi.org/10.1016/S1003-6326(09)60374-5
2. Li B, Wang HW, Jie JC, Wei ZJ (2011) Microstructure evolution and modification mechanism of the ytterbium modified Al–7.5%Si–0.045%Mg alloys. J Alloys Compd 509:3387–3392. https://doi.org/10.1016/j.jallcom.2010.12.081
3. Borodanski K, Selivorstov V, Dotsenko Y, Zinigrad M (2015) Effect of additions of ceramic nanoparticles and gas-dynamic treatment on Al alloys. Metals 5:2277–2288. https://doi.org/10.3390/met5042277
4. Wang H, Davidson CJ, St John DH (2004) Semisolid microstructural evolution of AlSi7 Mg alloy during partial remelting. Mater Sci Eng A 368:159–167. https://doi.org/10.1016/j.msea.2003.10.305
5. Prasada Rao AK, Das K, Murty BS (2008) Microstructural features of as-cast A356 alloy inoculated with Sr, Sb modifiers and Al–Ti–C grain refiner simultaneously. Mater Lett 62:273–275. https://doi.org/10.1016/j.matlet.2007.05.020
6. Makhlouf MM, Guthy HV (2001) The aluminum–silicon eutectic reaction: mechanisms and crystallography. J. Light Met 1:199–218. https://doi.org/10.1016/S1471-5317(02)00003-2
7. Shankar S, Riddle YW, Makhlouf MM (2004) Nucleation mechanism of the eutectic phases in aluminum–silicon hypoeutectic alloys. Acta Mater 52:4447–4460. https://doi.org/10.1016/j.actamat.2004.05.045
8. Timple M, Wanderka N, Schlesiger R, Yamamoto T, Lazarev N, Isheim D, Schmitz G, Matsumura S, Banhart J (2012) The role of strontium in modifying aluminium-silicon alloys. Acta Mater 60:3920–3928. https://doi.org/10.1016/j.actamat.2012.03.031
9. Salleh MS, Omar MZ, Syarif J (2015) The effects of Mg addition on the microstructure and mechanical properties of thixoformed Al-5%Si-Cu alloys. J Alloys Compd 621:121–130. https://doi.org/10.1016/j.jallcom.2014.09.152
10. Caceres CH, Davidson CJ, Griffiths JR, Wang QG (1999) The effect of Mg on the microstructure and mechanical behavior of Al-Si-Mg casting alloys. Metall Mater Trans A 30:2611–2618. https://doi.org/10.1007/s11661-999-0301-8
11. Dutta S, Allen SM (1991) A calorimetric study of precipitation in commercial aluminium alloy 6061. J Mater Sci Lett 10:323–326. https://doi.org/10.1007/BF00719697
12. Edwards GA, Stiller K, Dunlop GL, Couper MJ (1998) The precipitation sequence in Al–Mg–Si alloys. Acta Mater 46:3893–3904. https://doi.org/10.1016/S1359-6454(98)00059-7
13. Svoboda J, Fischer F, Fratzl P, Kozeschnik E (2004) Modelling of kinetics in multi-component multi-phase systems with spherical precipitates I: theory. J Mater Sci Eng A 385:166–174. https://doi.org/10.1016/S0921-5093(04)00820-2
14. Kavosi J, Seal M, Kazeminezhad M, Dodangeh A (2014) Modeling of dislocation density and strength on rheoforged A356 alloy during multi-directional forging. Comput Mater Sci 81:284–289. https://doi.org/10.1016/j.commatsci.2013.08.029
15. Wu L, Ferguson WG (2009) Modelling of precipitation hardening in casting aluminium alloys. Mater Sci Forum 618:203–206. https://doi.org/10.4028/www.scientific.net/MSF.618-619.203
16. Kozeschnik E, Svoboda J, Fratzl P, Fischer F (2004) Modelling of kinetics in multi-component multi-phase systems with spherical precipitates II: numerical solution and application. J Mater Sci Eng A 385:157–165. https://doi.org/10.1016/S0921-5093(04)00821-4
17. Falahati A, Ahmadi MR, Warczok P, Lang P, Povoden-Karadeniz E, Kozschnik E (2011) Thermo-kinetic computer simulation of precipitation and age-hardening effect in Al-Mg-Si alloys with arbitrary heat treatment. Mater Sci Technol 1:292–299
18. Ludwig A, Holzmann T (2016) Automatic optimization of localized heat treatment for Al-Si-Mg alloys. Mat Sci Eng 119:1–8. https://doi.org/10.1088/1757-899X/119/1/012026

19. Weller HG, Tabor G, Jasak H, Fureby C (1998) A tensorial approach to computational continuum mechanics using object-oriented techniques. Comput Phys 12:620–631. https://doi.org/10.1063/1.168744
20. Boley BA, Weiner JH (1985) Theory of thermal stresses. Robert E. Krieger Publishing Company, Huntington
21. Moukalled F, Mangani L, Darwish M (2016) The finite volume method in computational fluid dynamics: an advanced introduction with OpenFOAM and Matlab. Springer International Publishing. https://doi.org/10.1007/978-3-319-16874-6
22. Jasak H, Weller HG (2000) Application of the finite volume method and unstructured meshes to linear elasticity. Int J Numer Methods Eng 48:267–287
23. Fahrmeir L, Häußler W (1984) Multivariate statistische Verfahren. de Gruyter, Berlin, pp 20–37
24. Holzmann T (2017) C++ laser framework. http://www.bitbucket.org/shor-ty/laserconvectionbc

[illegible]

Thermodynamic Properties of Magnetic Semiconductors $Ag_2FeSn_3S_8$ and Ag_2FeSnS_4 Determined by the EMF Method

Mykola Moroz, Fiseha Tesfaye, Pavlo Demchenko, Myroslava Prokhorenko, Daniel Lindberg, Oleksandr Reshetnyak and Leena Hupa

Abstract Phase equilibria and thermodynamics in the Ag_2SnS_3–SnS–Sn_2S_3–FeS system were investigated using differential thermal analysis, X-ray diffraction, and EMF methods. Determined phase relations were used to express the forming chemical reactions for compounds $Ag_2FeSn_3S_8$ and Ag_2FeSnS_4. The forming chemical reactions were performed by applying electrochemical cells of the types: (−) C | Ag | Ag_3GeS_3I glass | $Ag_2FeSn_3S_8$, SnS, Sn_2S_3, FeS | C (+) and (−) C | Ag | Ag_3GeS_3I glass | Ag_2FeSnS_4, SnS, $Ag_2FeSn_3S_8$, FeS | C (+). Based on the measured EMF versus temperature relations, experimental thermodynamic data of the quaternary phases $Ag_2FeSn_3S_8$ and Ag_2FeSnS_4 were derived for the first time.

Keywords Magnetic semiconductors · Phase equilibria · Thermodynamic properties · EMF method · Gibbs energy

M. Moroz (✉) · F. Tesfaye · D. Lindberg · L. Hupa
Johan Gadolin Process Chemistry Centre, Åbo Akademi University,
20500 Turku, Finland
e-mail: riv018@i.ua

P. Demchenko
Department of Inorganic Chemistry, Ivan Franko National University of Lviv,
Lviv 79005, Ukraine

M. Prokhorenko
Department of Cartography and Geospatial Modeling, Lviv Polytechnic
National University, Lviv 79013, Ukraine

O. Reshetnyak
Department of Physical and Colloid Chemistry, Ivan Franko National
University of Lviv, Lviv 79005, Ukraine

G. Lambotte et al. (eds.), *Materials Processing Fundamentals 2018*,
The Minerals, Metals & Materials Series,
https://doi.org/10.1007/978-3-319-72131-6_8

Introduction

In the A–B–C–X systems (A = Ag, Cu, Li; B = Zn, Cd, Hg, Pb, Mn, Fe, Eu; C = Si, Ge, Sn; and X = S, Se, Te) the A_2BCX_4 compounds (I) are formed [1–16]. The acentric crystal structure of compounds (I) and wide spectral range of transparency windows make them promising materials for non-linear optics and optoelectronics applications [17]. Silver-containing compounds are also expected to be ionic (Ag^+) conductors, which make them interesting for electrochemical applications [18, 19]. Compounds (I) with B = Fe belong to the class of dilute magnetic semiconductors in which a fraction of cations are magnetic [10, 20]. The existence of magneto-transport and magneto-optical effects allows the use of these materials in magnetic field sensors as well as in magneto-optical devices [21]. In the case of B = (Mn, Fe) and X = (Se, Te), A_2BCX_4 decomposes below $T \approx 600$ K into two phases, while preserving the formula composition of the unit cells, but with different crystal structures [7, 10]. In the systems with A = (Ag, Cu); B = (Cd, Ni, Co, Fe, Mn); C = Sn; X = S compounds of the type $A_2BSn_3S_8$ (II) are formed [13, 22–25]. Compounds (II) with B = (Mn, Fe) are promising electrode materials for lithium batteries [24, 25]. The compositions of compounds (I) and (II) belong to the same plane of the quasi-ternary A_2X–BX–CX_2 systems.

In the Ag–Fe–Sn–S system two quaternary compounds are formed: $Ag_2FeSn_3S_8$ which belong to spinels constituting materials [26, 27], and Ag_2FeSnS_4 [28]. According to data of Yajima et al. [27], $Ag_2FeSn_3S_8$ (mineral Toyohaite) is crystallized in space group $I4_1/a$, $a = 7.464$, $c = 10.80$ Å, $Z = 2$. Garg et al. [24] obtained some different results when exploring $Ag_2FeSn_3S_8$ single crystals: space group $Fd\bar{3}m$, $a = 10.5723$ Å. The single crystals of $Ag_2FeSn_3S_8$ were obtained by heating of pure metals and sulfur at $T = 1023$ K and then cooling to room temperature. The compositions of the quaternary compounds belong to the Ag_2S–FeS–SnS_2 concentration plane. The intermediate phases Ag_8SnS_6 [29], Ag_2SnS_3 [30], $Ag_2Sn_2S_5$ [30], and $Ag_4Sn_3S_8$ [31] exist in the quasi binary Ag_2S–SnS_2 system. The compounds $Ag_{3.8}Sn_3S_8$ and $Ag_4Sn_3S_8$ are known to be superionic conductors [31, 32]. No intermediate phases have been reported in the literature in the binary Ag_2S–FeS and FeS–SnS_2 subsystems.

Investigation of the phase equilibria of the quasi-quaternary phase regions in the Ag–Fe–Sn–S system by differential thermal analysis (DTA) and X-ray diffraction (XRD) methods involves experimental difficulties. Incongruent melting of quaternary phases, low thermal stability, kinetic hindrances to polymorphic phase transformations, and change of the regions of stabilities of some phases with increasing temperature complicate the selection criteria to confirm the attainment of equilibrium state. These experimental difficulties can partially be solved through theoretical calculations of the T–x phase diagrams using, for example, the CALPHAD methods [33, 34]. These calculation methods are based on thermodynamic data of the pure phases in the system to be investigated together with information on T–x diagrams of separate sections. The thermodynamic properties of

Ag_2S, FeS, SnS, SnS_2, and Sn_2S_3 are reported in [35]. Thermodynamic data of Ag_8SnS_6, Ag_2SnS_3, and $Ag_2Sn_2S_5$ ($Ag_4Sn_3S_8$) are reported in [36].

In this work, we present the experimental results and calculation of the values of standard thermodynamic functions of saturated solid solutions of the phases $Ag_2FeSn_3S_8$ and Ag_2FeSnS_4 in the quaternary Ag–Fe–Sn–S system. The Gibbs energies, enthalpies, and entropies of formations of the phases were determined by applying the electromotive force (EMF) method [36, 37]. The data on the thermodynamic properties of the quaternary phases can be used for thermodynamic modeling of the unknown phase regions of the Ag–Fe–Sn–S system.

Experimental

The phases were synthesized from high-purity elements (99.99 wt% Ag, 99.9 wt % Fe, 99.999 wt% Sn, and 99.99 wt% S). The synthesis and annealing were performed in thin-walled evacuated quartz glass ampoules with a residual pressure ≤ 1 Pa. Compounds SnS, Sn_2S_3, FeS, and Ag_2SnS_3 were synthesized by cooling the respective melts of elements from $T = 1200$ K. Compounds Ag_2FeSnS_4 and $Ag_2FeSn_3S_8$ were obtained through solid-state synthesis from finely-disperse mixtures of Ag_2SnS_3 and FeS in the molar ratio 1 : 1, and Ag_2SnS_3, FeS, and SnS_2 in the molar ratio 1 : 1 : 2, respectively, at $T = 950$ K for 200 h. The high-temperature annealing of quaternary compounds was performed at 750 K for 240 h. DTA and XRD methods were used to characterize the phase composition of the synthesized compounds.

The DTA curves of the samples were recorded using a Paulik-Paulik-Erdey derivatograph (Hungary) fitted with chromel-alumel thermocouples and an H307-1 XY recorder (Ukraine). The heating and cooling rates in the DTA measurements of the samples were in the range 6–8 K min^{-1}. The thermocouples were calibrated by the melting temperatures of In ($T = 429$ K), Sn ($T = 505$ K), Cd ($T = 594$ K), Te ($T = 723$ K), Sb ($T = 904$ K), NaCl ($T = 1074$ K), Ge ($T = 1209$ K), Ag ($T = 1236$ K), and Cu ($T = 1357$ K). Errors in the temperature measurements were below $\Delta T = \pm 3$ K. XRD patterns were collected on a STOE STADI P diffractometer equipped with a linear position–sensitive detector PSD, in a modified Guinier geometry (transmission mode, $CuK\alpha_1$ radiation, a bent Ge (111) monochromator, $2\theta/\omega$ scan mode). Preliminary data processing and X-ray phase analyses were performed using STOE WinXPOW 3.03 [38] and PowderCell 2.4 PC programs [39], using data on crystal structures for the phases taken from database [40].

Potential-forming processes were performed by applying the following electrochemical cells (ECCs):

$$(-)\text{C} \mid \text{Ag} \mid \text{Ag}_3\text{GeS}_3\text{I glass} \mid \text{Ag}_2\text{FeSn}_3\text{S}_8,\ \text{SnS},\ \text{Sn}_2\text{S}_3,\ \text{FeS} \mid \text{C}(+), \quad \text{(i)}$$

$$(-)\text{C} \mid \text{Ag} \mid \text{Ag}_3\text{GeS}_3\text{I glass} \mid \text{Ag}_2\text{FeSnS}_4,\ \text{SnS},\ \text{Ag}_2\text{FeSn}_3\text{S}_8,\ \text{FeS} \mid \text{C}(+), \quad \text{(ii)}$$

where C is graphite and Ag_3GeS_3I glass is the fast purely Ag^+ ions conducting electrolyte [41]. Ag_3GeS_3I glass has similar ionic properties with superionic materials Ag_2GeS_3 and Ag_3GeS_3Br [42, 43]. The positive electrodes in the ECCs (i) and (ii) were prepared by melting mixtures of (Sn_2S_3 : FeS : $Ag_2FeSn_3S_8$: SnS) in the molar ratio 4 : 1 : 1 : 5 and ($Ag_2FeSn_3S_8$: FeS : Ag_2FeSnS_4 : SnS) in the molar ratio 1 : 1 : 2 : 1, respectively. The equilibrium state of four-phases sample was achieved by vacuum annealing at 750 K for 240 h. Components of the ECCs in powder form were pressed at 10^8 Pa through a 2 mm diameter hole arranged in the fluoroplast matrix up to density $\rho = (0.93 \pm 0.02)\cdot\rho_0$, where ρ_0 is the experimentally determined density of cast samples. To eliminate possible defects of plastic deformation during the pressing of the samples, we performed five-fold thermal cycling of ECCs in the range of 400–550 K, with heating and cooling rates of 2 K min^{-1}. The ECCs were heated in a resistance furnace, similar to that described in [44], filled with a mixture of H_2 and Ar (both 0.9999 volume fraction) in the molar ratio 1 : 9, with $P = 1.2 \times 10^5$ Pa. Highly pure hydrogen was obtained by the diffusion of the bottled gas through palladium foil. Argon gas was purified from traces of oxygen in a quartz glass tube by passing it through copper foil heated at 673 K. The flow of gas at the rate of 2×10^{-3} m^3 h^{-1} had the direction from the positive to the negative electrodes of the ECCs. The temperature was maintained with an accuracy of ± 0.5 K. The EMF values of the cells were measured using the voltmeter U7–9 electrometric amplifier (Ukraine) which has an input impedance of $>10^{12}$ Ω. The equilibrium in ECCs at each temperature was achieved within 2 h. The equilibrium was considered to have been reached when the EMF-values were constant or their variations were not significant (below $\Delta E = \pm$ 0.2 mV). The dependences of the EMF of the cells on temperature $E(T)$ were analyzed by the method described in [45].

Results and Discussion

Triangulation of the Ag–Fe–Sn–S system below T = 570 K in the Ag_2SnS_3–SnS–Sn_2S_3–FeS part that includes the compounds $Ag_2FeSn_3S_8$ and Ag_2FeSnS_4 is shown in Fig. 1.

The two-phase equilibria in the FeS–Sn_2S_3, FeS–SnS, $Ag_2FeSn_3S_8$–Sn_2S_3, $Ag_2FeSn_3S_8$–SnS, and Ag_2FeSnS_4–SnS systems were determined in this work by XRD, DTA, and EMF methods. As an example, the two-phase state of the $Ag_2FeSn_3S_8$–Ag_2FeSnS_4 and $Ag_2GeSn_3S_8$–SnS sections are supported by the diffraction patterns presented in Fig. 2. The formation of the compound Sn_3S_4 [46] was not confirmed in our investigation. The diffraction pattern of "Sn_3S_4" contains only reflections of SnS and Sn_2S_3 compounds.

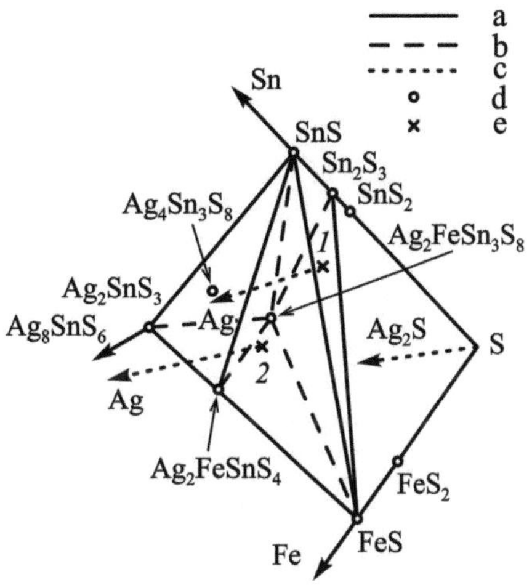

Fig. 1 Triangulation of Ag–Fe–Sn–S system below $T = 570$ K in the Ag_2SnS_3–SnS–Sn_2S_3–FeS part: a and b are the lines of two-phase equilibria, c represents sections of concentration space, d represents figurative points of compounds, e represents compositions of positive electrodes of phase regions $Ag_2FeSn_3S_8$–SnS–Sn_2S_3–FeS (1) and Ag_2FeSnS_4–SnS–$Ag_2FeSn_3S_8$–FeS (2)

The phase relations presented in Fig. 1 were used to write the equations of the potential forming chemical reaction. The electrochemical process of the formation of $Ag_2FeSn_3S_8$ from silver and compounds can be written as follows:

$2Ag = 2Ag^+ + 2e^-$ left side electrode (reference system),

$2Ag^+ + 2e^- + 4Sn_2S_3 + FeS = Ag_2FeSn_3S_8 + 5SnS$ right side electrode (sample system),

$2Ag + 4Sn_2S_3 + FeS = Ag_2FeSn_3S_8 + 5SnS$ overall cell reaction. (1)

The electrochemical process of the formation of Ag_2FeSnS_4 from silver and compounds can be written as follows:

$2Ag = 2Ag^+ + 2e^-$ left side electrode (reference system),

$2Ag^+ + 2e^- + Ag_2FeSn_3S_8 + FeS = 2Ag_2FeSnS_4 + SnS$ right side electrode (sample system),

$2Ag + Ag_2FeSn_3S_8 + FeS = 2Ag_2FeSnS_4 + SnS$ overall cell reaction. (2)

Thus, the overall cell reactions (1) and (2) took place in ECCs (i) and (ii), respectively. Reactions (1) and (2) were expressed by assuming small differences in the positions of the figurative points of the compounds and their saturated solid solutions in the Gibbs tetrahedron.

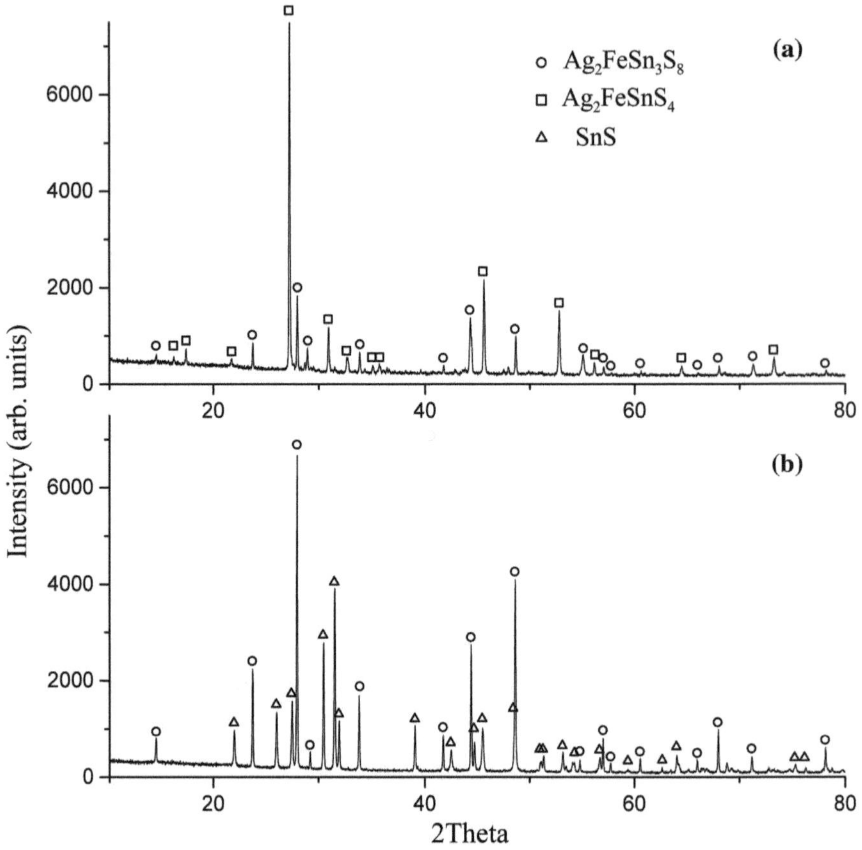

Fig. 2 Diffraction patterns for the analyzed phase equilibria on the tie lines of the Ag_2SnS_3–SnS–Sn_2S_3–FeS system; **a** $Ag_2FeSn_3S_8$–Ag_2FeSnS_4 and **b** $Ag_2FeSn_3S_8$–SnS

The temperature dependences of the EMF of the cells (i) and (ii) are shown in Fig. 3. The relationship of E versus T measured with ECCs (i) and (ii) were approximated by Eqs. 3 and 4, respectively:

$$E_{(1)}/\mathrm{mV} = (256.73 \pm 0.46) + (99.04 \pm 0.93) \times 10^{-3} \quad T/K \quad 450 \le T/K \le 538, \tag{3}$$

$$E_{(2)}/\mathrm{mV} = (-79.67 \pm 1.98) + (682.73 \pm 3.99) \times 10^{-3} \; T/K \quad 452 \le T/K \le 538. \tag{4}$$

Gibbs energy of the reaction (1) and (2) can be calculated from the measured E versus T relations by using the fundamental thermodynamic equation:

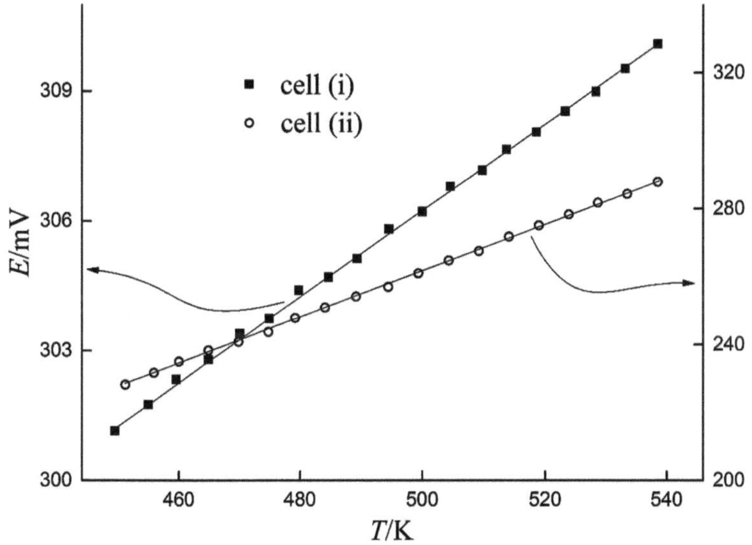

Fig. 3 The measured E versus T relations of ECCs (i) and (ii)

$$\Delta_r G = -z \cdot F \cdot E, \tag{5}$$

where $z = 2$ is the number of electrons involved in the electrochemical cell reactions (1) and (2), $F = 96{,}485.3\ \text{C}\cdot\text{mol}^{-1}$ is the Faraday constant, and E is the EMF produced by electrochemical cell.

The Gibbs energy of the reactions (1) and (2) are related to the Gibbs energies of formations of the compounds according to Eqs. (6) and (7), respectively.

$$\Delta_{r(1)}G^{\circ} = \Delta_f G^{\circ}_{Ag_2FeSn_3S_8} + 5\Delta_f G^{\circ}_{SnS} - 4\Delta_f G^{\circ}_{Sn_2S_3} - \Delta_f G^{\circ}_{FeS}, \tag{6}$$

$$\Delta_{r(2)}G^{\circ} = 2\Delta_f G^{\circ}_{Ag_2FeSnS_4} + \Delta_f G^{\circ}_{SnS} - \Delta_f G^{\circ}_{Ag_2FeSn_3S_8} - \Delta_f G^{\circ}_{FeS}. \tag{7}$$

Equations (8) and (9) were obtained from Eqs. (6) and (7), respectively.

$$\Delta_f G^{\circ}_{Ag_2FeSn_3S_8} = \Delta_{r(1)}G^{\circ} + 4\Delta_f G^{\circ}_{Sn_2S_3} + \Delta_f G^{\circ}_{FeS} - 5\Delta_f G^{\circ}_{SnS}, \tag{8}$$

$$\Delta_f G^{\circ}_{Ag_2FeSnS_4} = 0.5\left[\Delta_{r(2)}G^{\circ} + \Delta_f G_{Ag_2FeSn_3S_8} + \Delta_f G^{\circ}_{FeS} - \Delta_f G^{\circ}_{SnS}\right]. \tag{9}$$

Using a linear least square fitting of the thermodynamic data presented in the handbook of Barin et al. [35], Gibbs energies of formations of SnS, Sn_2S_3, and FeS compounds can be expressed as:

$$\Delta_f G^\circ_{SnS}/(\mathrm{kJ\,mol^{-1}}) = -(114.60 \pm 2.40) + (22.49 \pm 4.73)10^{-3}\ T/\mathrm{K} \quad 400 \leq T/\mathrm{K} \leq 600, \tag{10}$$

$$\Delta_f G^\circ_{Sn_2S_3}/(\mathrm{kJ\,mol^{-1}}) = -(280.50 \pm 5.26) + (75.71 \pm 10.37)10^{-3}\ T/\mathrm{K} \quad 400 \leq T/\mathrm{K} \leq 600, \tag{11}$$

$$\Delta_f G^\circ_{FeS}/(\mathrm{kJ\,mol^{-1}}) = -(99.95 \pm 0.10) - (5.21 \pm 0.20)10^{-3}\ T/\mathrm{K} \quad 411 \leq T/\mathrm{K} \leq 598. \tag{12}$$

By combining Eqs. (5), (8), and (9) with Eqs. (10)–(12), the standard Gibbs energies of formations of $Ag_2FeSn_3S_8$ and Ag_2FeSnS_4 were calculated to be:

$$\Delta_f G^\circ_{Ag_2FeSn_3S_8}/\left(\mathrm{kJ\,mol^{-1}}\right) = -(698.49 \pm 24.22) + (166.07 \pm 47.75) \times 10^{-3}\ T/\mathrm{K} \quad 450 \leq T/\mathrm{K} \leq 538, \tag{13}$$

$$\Delta_f G^\circ_{Ag_2FeSnS_4}/\left(\mathrm{kJ\,mol^{-1}}\right) = -(334.23 \pm 12.17) + (3.31 \pm 23.99) \times 10^{-3}\ T/\mathrm{K} \quad 452 \leq T/\mathrm{K} \leq 538. \tag{14}$$

Since FeS undergoes a phase transition at 411 K [35], the Gibbs energies of phase transition was considered at temperatures below 411 K. According to basic thermodynamics, Gibbs energy of phase transition at the transformation temperature can be expressed by [47]:

$$\Delta_{tr} G^\circ = \Delta_{tr} H^\circ - \left(\frac{\Delta_{tr} H^\circ}{T_{tr}}\right)\Delta T. \tag{15}$$

By using the enthalpy of phase transition given in the handbook [35]:

$$\alpha - \mathrm{FeS} \leftrightarrow \beta - \mathrm{FeS} \quad (\Delta_{tr} H^\circ_{FeS} = 4.051\ \mathrm{kJ\,mol^{-1}}), \tag{16}$$

the high temperature experimental data, expressed by Eqs. (13) and (14), were extrapolated to lower temperatures with α-FeS as standard state by combining Eqs. (15) and (16). The least square analysis yields:

$$\Delta_f G^\circ_{Ag_2FeSn_3S_8}/\left(\mathrm{kJ\,mol^{-1}}\right) = -(694.44 \pm 24.22) + (156.21 \pm 47.75) \times 10^{-3}\ T/\mathrm{K} \quad 298 \leq T/\mathrm{K} \leq 411, \tag{17}$$

Table 1 Standard thermodynamic properties of the selected phases in the Ag–Fe–Sn–S system at 298 K and 1 atm

Phase	$-\Delta_f G^\circ$	$-\Delta_f H^\circ$	S°	[Ref.] Note
	kJ·mol$^{\circ-1}$		J·mol$^{\circ-1}$·K$^{\circ-1}$	
Ag	0	0	42.677	[35]
Fe	0	0	27.280	[35]
Sn	0	0	51.195	[35]
S	0	0	32.056	[35]
SnS	106.079	107.947	76.986	[35]
Sn_2S_3	253.417	263.592	164.431	[35]
FeS	101.965	101.671	60.321	[35]
$Ag_2FeSn_3S_8$	647.87 ± 28.10	694.44 ± 24.22	678.88 ± 47.75	This work
Ag_2FeSnS_4	332.14 ± 14.12	330.18 ± 12.17	285.51 ± 23.99	This work

$$\Delta_f G^\circ_{Ag_2FeSnS_4}/(\text{kJ mol}^{-1}) = -(330.18 \pm 12.17) - (6.54 \pm 23.99) \times 10^{-3}\ T/\text{K} \quad 298 \le T/\text{K} \le 411. \tag{18}$$

The standard entropy of the quaternary compounds $S^\circ_{Ag_2FeSn_3S_8}$ and $S^\circ_{Ag_2FeSnS_4}$ at 298 K can be derived from the $\Delta_f S^\circ_{Ag_2FeSn_3S_8}$ and $\Delta_f S^\circ_{Ag_2FeSnS_4}$ values of Eqs. (17) and (18) by using Eqs. (19) and (20):

$$S^\circ_{Ag_2FeSn_3S_8} = \Delta_f S^\circ_{Ag_2FeSn_3S_8} + 2S^\circ_{Ag} + S^\circ_{Fe} + 3S^\circ_{Sn} + 8S^\circ_{S}, \tag{19}$$

$$S^\circ_{Ag_2FeSnS_4} = \Delta_f S^\circ_{Ag_2FeSnS_4} + 2S^\circ_{Ag} + S^\circ_{Fe} + S^\circ_{Sn} + 4S^\circ_{S}. \tag{20}$$

By applying Eqs. (17)–(20), standard Gibbs energy, enthalpy and entropy of formations of $Ag_2FeSn_3S_8$ and Ag_2FeSnS_4 at 298 K were calculated. A comparative summary of the calculated values together with the available literature values is presented in Table 1.

Summary and Conclusions

1. Phase equilibria of the Ag–Fe–Sn–S system in the Ag_2SnS_3–SnS–Sn_2S_3–FeS part were established by applying XRD, DTA and EMF methods.
2. The measured EMF versus temperature values were used to calculate the Gibbs energies of reactions (1) and (2) as well as Gibbs energies of formations of the quaternary phases $Ag_2FeSn_3S_8$ and Ag_2FeSnS_4 for the first time as:

$$\Delta_f G^{\circ}_{Ag_2FeSn_3S_8}\left(\mathrm{kJ\,mol^{-1}}\right) = -(694.44 \pm 24.22) + (156.21 \pm 47.75) \times 10^{-3} T/K$$
$$298 \leq T/K \leq 411,$$

$$\Delta_f G_{Ag_2FeSnS_4}/\left(\mathrm{kJ\ mol^{-1}}\right) = -(330.18 \pm 12.17) - (6.54 \pm 23.99) \times 10^{-3}\ T/\mathrm{K}$$
$$298 \leq T/K \leq 411.$$

3. The standard thermodynamic functions of the phases determined in this work can be used for thermodynamic modeling of the *T*–*x* diagrams of the Ag–Fe–Sn–S system.

Acknowledgements The authors are grateful to the Johan Gadolin Scholarship programme and Academy of Finland for financial support. This work was made under the project "Thermodynamic investigation of complex inorganic material systems for improved renewable energy and metals production processes" as part of the activities of the Johan Gadolin Process Chemistry Center at Åbo Akademi University.

References

1. Parthé E, Yvon K, Deitch RH (1969) The crystal structure of Cu2CdGeS4 and other quaternary normal tetrahedral structure compounds. Acta Crystallogr B 25:1164–1174
2. Schäfer W, Nitsche R (1974) Tetrahedral quaternary chalcogenides of the type Cu2-II-IV-S4 (Se4). Mater Res Bull 9:645–654
3. Lamarche G, Woolley JC, Tovar R, Quintero M, Sagredo V (1989) Effects of crystallographic ordering on the magnetic behaviour of (AgIn)1-zMn2zTe2 and (CuIn)1-zMn2zTe2 alloys. J Magn Magn Mater 80:321–328
4. Lamarche A-M, Willsher A, Chen L, Lamarche G, Woolley JC (1991) Crystal structures of I2·Mn·IV·VI4 compounds. J Solid State Chem 94:313–318
5. Chen XL, Lamarche A-M, Lamarche G, Woolley JC (1993) Magnetic behaviour of some I2·Mn·IV·VI4 compounds. J Magn Magn Mater 118:119–128
6. Chen XL, Lamarche A-M, Lamarche G, Woolley JC (1993) Effects of magnetic transitions on the optical energy gap values of some I2.Mn.IV.VI4 phases. J Phys Condens Matter 5:7143–7154
7. Woolley JC, Lamarche A-M, Lamarche G, Church C, Swainson IP, Holden TM (1995) Crystal symmetry of Ag2MnGeTe4 phases. J Solid State Chem 115:192–196
8. Marking GA, Hanko JA, Kanatzidis MG (1998) New quaternary thiostannates and thiogermanates A2Hg3M2S8 (A = Cs, Rb; M = Sn, Ge) through Molten A2Sx. Reversible Glass Formation in Cs2Hg3M2S8. Chem Mater 10:1191–1199
9. Aitken JA, Larson P, Mahanti SD, Kanatzidis MG (2001) Li2PbGeS4 and Li2EuGeS4: polar chalcopyrites with a severe tetragonal compression. Chem Mater 13:4714–4721
10. Quintero M, Barreto A, Grima P, Tovar R, Quintero E, Porras GS, Ruiz J, Woolley JC, Lamarche G, Lamarche A-M (1999) Crystallographic properties of I2–Fe–IV–VI4 magnetic semiconductor compounds. Mater Res Bull 34:2263–2270
11. Parasyuk OV, Gulay LD, Piskach LV, Gagalovska OP (2002) The Ag2S–HgS–GeS2 system at 670 K and the crystal structure of the Ag2HgGeS4 compound. J Alloys Compd 336:213–217

12. Parasyuk OV, Piskach LV, Romanyuk YE, Olekseyuk ID, Zaremba VI, Pekhnyo VI (2005) Phase relations in the quasi-binary Cu2GeS3–ZnS and quasi-ternary Cu2S–Zn(Cd)S–GeS2 systems and crystal structure of Cu2ZnGeS4. J Alloys Compd 397:85–94
13. Parasyuk OV, Olekseyuk ID, Piskach LV, Volkov SV, Pekhnyo VI (2005) Phase relations in the Ag2S–CdS–SnS2 system and the crystal structure of the compounds. J Alloys Compd 399:173–177
14. Parasyuk OV, Fedorchuk AO, Kogut YM, Piskach LV, Olekseyuk ID (2010) The Ag2S–ZnS–GeS2 system: Phase diagram, glass-formation region and crystal structure of Ag2ZnGeS4. J Alloys Compd 500:26–29
15. Kogut Y, Fedorchuk A, Zhbankov O, Romanyuk Y, Kityk I, Piskach L, Parasyuk O (2011) Isothermal section of the Ag2S–PbS–GeS2 system at 300K and the crystal structure of Ag2PbGeS4. J Alloys Compd 509:4264–4267
16. Greil S(2016) Untersuchungen an ternären und quaternären Kupfer-, Lithium-, und Silbersulfiden mit Diamantstruktur, PhD, Universität Regensburg
17. Davydyuk GE, Myronchuk GL, Kityk IV, Danyl'chuk SP, Bozhko VV, Parasyuk OV (2011) Ag2CdSnS4 single crystals as promising materials for optoelectronic. Opt Mater 33:1302–1306
18. Parasyuk OV, Gulay LD, Romanyuk YE, Olekseyuk ID (2003) The Ag2Se–HgSe–SiSe2 system in the 0–60 mol.% SiSe2 region. J Alloys Compd 348:157–166
19. Parasyuk OV, Gulay LD, Romanyuk YE, Olekseyuk ID, Piskach LV (2003) The Ag2Se–HgSe–GeSe2 system and crystal structures of the compounds. J Alloys Compd 351:135–144
20. Averous M, Balkanski M (2012) Semimagnetic semiconductors and diluted magnetic semiconductors. Springer Science & Business Media
21. Dietl T (2010) A ten-year perspective on dilute magnetic semiconductors and oxides. Nat Mater 9:965–974
22. Jumas JC, Philippot E, Maurin M (1979) Structure du rhodostannite synthétique. Acta Crystallogr B 35:2195–2197
23. Chykhrij S, Sysa L, Parasyuk O, Piskach L (2000) Crystal structure of the Cu2CdSn3S8 compound. J Alloys Compd 307:124–126
24. Garg G, Gupta S, Maddanimath T, Gascoin F, Ganguli AK (2003) Single crystal structure, electrical and electrochemical properties of the quaternary thiospinel: Ag2FeSn3S8. Solid State Ionics 164:205–209
25. Garg G, Ramanujachary KV, Lofland SE, Lobanov MV, Greenblatt M, Maddanimath T, Vijayamohanan K, Ganguli AK (2003) Crystal structure, magnetic and electrochemical properties of a quaternary thiospinel: Ag2MnSn3S8. J Solid State Chem 174:229–232
26. Padiou J, Jumas JC, Ribes M (1981) Sur une nouvelle famille de composes quaternaires MM'0.5Sn1.5S4 (M=Cu, Ag; M'= Mn, Fe Co, Ni) de type spinelle: caracterisation par spectroscopie Mossbauer de 119Sn et proprietes magnetiques. Rev Chim Miner 18:33–42
27. Yajima J, Ohta E, Kanazawa Y (1991) Toyohaite, Ag2FeSn3S8, a new mineral. Mineral J 15:222–232
28. Caye R, Laurent Y, Picot P, Pierrot R, Levy C (1968) La hocartite, Ag2SnFeS4, une nouvelle espece minerale. Bull Soc Fr Miner Cristal 91:383–387
29. Gorochov O (1968) Les composes Ag8MX6 (M=Si, Ge, Sn et X=S, Se, Te). Bull Soc Chim Fr 6:2263–2275
30. Kokhan OP (1996) Phase relations in the systems Ag2X–BIVX2 (BIV – Si, Ge, Sn; X – S, Se) and properties of compounds. PhD thesis, Uzhgorod State University (in Ukrainian)
31. Hull S, Berastegui P, Grippa A (2005) Ag+ diffusion within the rock-salt structured superionic conductor Ag4Sn3S8. J Phys Condens Matter 17:1067–1084
32. Amiel O, Frankel DC, Wada H (1995) Crystal structure and conductivity of the new superionic conductors Ag4Zr3S8 and Ag3.8Sn3S8. J Solid State Chem 116:409–421
33. Ipser H, Mikula A, Katayama I (2010) Overview: The emf method as a source of experimental thermodynamic data. Calphad 34:271–278
34. Kroupa A (2013) Modelling of phase diagrams and thermodynamic properties using Calphad method—development of thermodynamic databases. Comput Mater Sci 66:3–13

35. Barin I (1995) Thermochemical data of pure substance. VCH, Weinheim
36. Babanly MB, Yusibov YA, Babanly NB (2011) The EMF method with solid-state electrolyte in the thermodynamic investigation of ternary copper and silver chalcogenides. In: S Kara (Ed), InTech
37. Morachevskii AG, Voronin GF, eyderikh VA, Kutsenyuk IB (2003) Electrochemical methods investigations in thermodynamics of metallic system. Akademkniga (in Russian)
38. Diffractom (2010) Stoe WinXPOW (version 3.03), Stoe Cie GmbH Darmstadt
39. Kraus W, Nolze G (1996) POWDER CELL—program for the representation and manipulation of crystal structures and calculation of the resulting X-ray powder patterns. J Appl Crystallogr 29:301–303
40. Villars P, Cenzual E (eds) (2014) Pearson's crystal data: crystal structure database for inorganic compounds, release 2014/15. ASM International, Materials Park, Ohio, USA
41. Robinel E, Kone A, Duclot MJ, Souquet JL (1983) Silver sulfide based glasses:(II). Electrochemical properties of GeS2-Ag2S-Agl glasses: transference number measurement and redox stability range. J Non-Cryst Solids 57:59–70
42. Moroz MV, Prokhorenko MV, Demchenko PY, Reshetnyak OV (2017) Thermodynamic properties of saturated solid solutions of Ag7SnSe5Br and Ag8SnSe6 compounds in the Ag–Sn–Se–Br system measured by the EMF method. J Chem Thermodyn 106:228–231
43. Moroz MV, Prokhorenko MV, Reshetnyak OV, Demchenko PY (2017) Electrochemical determination of thermodynamic properties of saturated solid solutions of Hg2GeSe3, Hg2GeSe4, Ag2Hg3GeSe6, and Ag1.4Hg1.3GeSe6 compounds in the Ag–Hg–Ge–Se system. J Solid State Electrochem 21:833–837
44. Tesfaye F, Taskinen P (2014) Electrochemical study of the thermodynamic properties of matildite (β-AgBiS2) in different temperature and compositional ranges. J Solid State Electrochem 18:1683–1694
45. Osadchii EG, Echmaeva EA (2007) The system Ag-Au-Se: Phase relations below 405 K and determination of standard thermodynamic properties of selenides by solid-state galvanic cell technique. Am Mineral 92:640–647
46. Albers W, Schol K (1961) The P-T-X phase diagram of the system Sn-S. Philips Res Rep 16:329–342
47. Tesfaye F, Taskinen P (2014) Experimentally determined thermodynamic properties of schapbachite (α-AgBiS2) below T=700 K. J Chem Thermodyn 70:219–226

Effects of Heat Treatment on the Electrochemical Performance of Al Based Anode Materials for Air-Battery

Xingyu Gao, Jilai Xue, Xuan Liu and Gaojie Shi

Abstract Aluminum based alloy can serve as anode materials for air battery that has great potential for use in electrical vehicles and other green energy applications. In this paper, effects of heat-treating Al-0.2 Mg-0.2 Ga-0.4In-0.15Sn anode materials on their electrochemical performance and corrosion behavior have been investigated. The results show that the discharge voltage of the as-cast alloy anode at 20 $mAcm^{-2}$ is higher than that of the pure Al (about 300 mV on average) in 4 M KOH solution. In general, electrochemical performance is improved with increased temperature and processing time in heat treatment. The high energy density (about 3180 $Whkg^{-1}$) and discharge voltage (about 1.62 V) can be obtained with heat treatment at 550 °C for 12. SEM observation reveals that the micro cracks due to the corrosion around the grain boundaries of the Al alloy anodes appear after the discharge performance in 4 M KOH solution, which may reduce the charge transfer resistance crossing the reaction layer on the anode surface.

Keywords Aluminum alloy · Air battery · Heat treatment · Discharge curve

Introduction

Metal air batteries that can convert the chemical energy of metal into electricity directly with oxygen in the air have received great attention due to their technical advantages, such as their sustainability, safety and environmental-friendly nature. Aluminum, among others, with its high energy density (8100 $Whkg^{-1}$) and low cost, has been one of the most promising candidates as the anode material for metal air batteries [1–3]. As a result of expansion in consumption of aluminum products, the potential demand for aluminum anode materials in metal-air battery may help to compensate, to some extent, the overcapacity nowadays in aluminum smelters.

X. Gao · J. Xue (✉) · X. Liu · G. Shi
School of Metallurgical and Ecological Engineering, University of Science and Technology Beijing, Xueyuan Road 30, Beijing 100083, China
e-mail: jx@ustb.edu.cn

G. Lambotte et al. (eds.), *Materials Processing Fundamentals 2018*,
The Minerals, Metals & Materials Series,
https://doi.org/10.1007/978-3-319-72131-6_9

However, some serious problems limit the development of Al-air battery, such as the low anode efficiency and the large polarization. During the battery operation, there was a layer of protective oxide layer spontaneously formed on Al anode surface [4], which reduced the energy density and limit the power output of the Al-air batteries. To solve this problem, various technical approaches have been tried, for instances, using alkaline electrolytes [5] and corrosion inhibitors in electrolytes [6] or alloy elements such as Mg, Ga, In, Sn, etc. [7–9] in Al anodes and heat treatment of the alloy. The purpose of these research efforts was to reduce the self-corrosion of Al and enhance its electrochemical activity [10]. The addition of magnesium to aluminum improves its impurity intolerance, because Mg could form the compounds with elements like silicon, reducing cathodic sites for hydrogen evolution. The alloying elements, Ga, In, Sn, have lower melting temperature and high hydrogen overpotential, which are nobler than aluminum and more soluble in alkaline electrolytes. In general, casting and heat treatment process are the common industrial practice in make aluminum products [11]. However, their influence on the electrochemical performance of Al anode is seldom reported.

In this paper, both of the alloying elements and heating treatment have been investigated. The Al-0.2 Mg-0.2 Ga-0.4In-0.15Sn alloy was subject to different heating treatments in order to improve the electrochemical activity and polarization resistance for use as the anode Materials.

Materials and Methods

Material Preparation

Raw materials are high pure Al (99.999), Mg (99.9%), Ga (99.9%), In (99.99%), Sn (99.5%) metals for casting anode alloy. They were melted in a magnesia crucible in WZG-2 kg vacuum induction furnace under argon atmosphere at 780 ± 5 °C. Then the molten alloy was poured into a preheated cast iron dye. The ultimate composition of the alloys was analyzed by ICP-OES. The alloy was Al-0.2 Mg-0.2 Ga-0.4In-0.15Sn, while the high purity Al (99.999%) was used as a reference for comparison. After casting, the alloys were heat-treated (Table 1), and then were quenched in cold water. The samples after heat treatment were cut and machined with an exposed area of 1 cm^2. The exposed area of anode sample was ground with emery paper (grade from 200, 600, 1000 to 2000) and cleaned with distilled water.

Table 1 The heat treatment of Al based alloy samples

Temperature (°C)	Time		
	6 h	9 h	12 h
450	450 °C-6 h	450 °C-9 h	450 °C-12 h
500	500 °C-6 h	500 °C-9 h	500 °C-12 h
550	550 °C-6 h	550 °C-9 h	550 °C-12 h

Electrochemical Measurements

The electrochemical tests were carried out with three-electrode system using CHI660C electrochemical test system (CHI Company, USA). A platinum sheet served as the counter electrode and a mercury/mercuric oxide (Hg/HgO) was used as the reference electrode. The solution used in this study was 4 M KOH. The open-circuit potential (OCP) of the anode samples was tested first for 2 h to ensure that the electrode was stable. The potentiodynamic cyclic polariztion was performed at a scan rate of 1 mVs^{-1} after OCP measurement. The scanning scope of the anodic polarizing curve was from −2.5 to −0.5 V. The electrochemical impedance spectroscopy (EIS) measurements were carried out at OCP with a 5 mV sine perturbation. The measuring frequency range was from 100 kHz to 0.1 Hz.

Battery Performance Test

The batteries consisted of anode, cathode and electrolyte, where anodes were pure Al or Al-0.2 Mg-0.2 Ga-0.4In-0.15Sn alloy. The cathode was the air electrode with MnO_2 catalyst, and electrolytes were 4 M KOH solution. The discharge performance of Al-air batteries was studied by means of constant current discharge test at current densities of 20 $mAcm^{-2}$ for 5 h. The Land (a trademark for product from Land Co, China) battery test system was used to measure the discharge performance.

The weight of the anode was measured both before and after discharge test. The energy density for the tested battery system was calculated using the following formulas:

$$\text{Energy density} = UIT/\Delta m \tag{1}$$

where U is the average voltage, v; I is the current, mA; T is the time, h; Δm is the weight loss, g.

Corrosion Morphology After Discharge

After the discharge test, the anode samples were clean-out in solutions of 2% CrO_3 and 5% H_3PO_4 at 80 °C for about 5 min, and then rinsed by distilled water and ethanol. Their surface micrographs were examined using JSM6480LV scanning electro-microscope (SEM).

Results and Discussion

Polarization and Impedance Spectroscopy

Figure 1 is the Open-circuit potential (OCP) of pure Al and alloy anodes versus time. It is clearly that the OCP of the alloy anode is more negative than pure Al around 370 mV.

Figure 2 and Table 2 present the potentiodynamic polarization curve and corresponding corrosion parameters of the samples. The corrosion potential of alloy (−1.751 V) is more negative than pure Al (−1.364 V). Combined with the OCP test, it means that the addition of Mg, Ga, In, Sn elements can improve the electrochemical activity in 4 M KOH solution. The corrosion current density (I_{corr}) of the alloy anode was lower than the pure Al. The elements can also reduce the corrosion rate of the Al anode. The alloy had low negative potential and

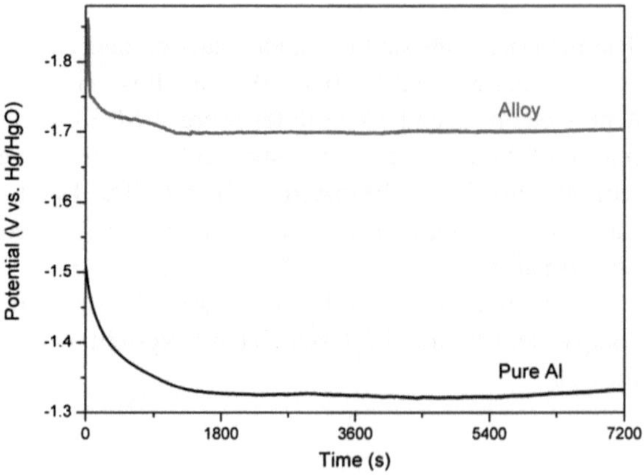

Fig. 1 Open-circuit potential versus time for pure Al and alloy anodes in 4 M KOH solution for 2 h

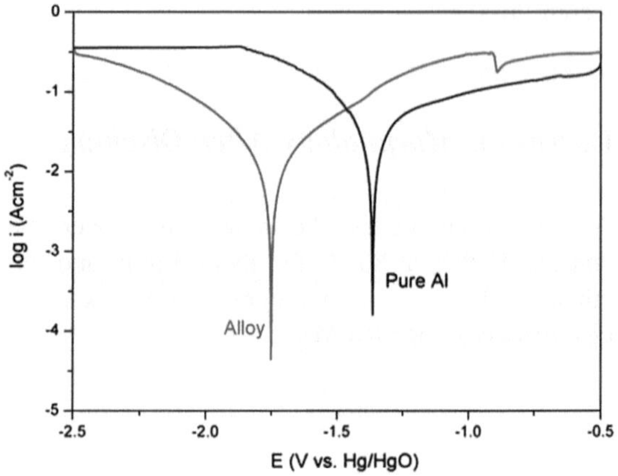

Fig. 2 Potentiodynamic polarization curves for pure alloy anode in 4 M KOH solution at a scan rate of 1 mVs^{-1}

Table 2 Corrosion parameters of different anode materials in 4 M KOH

Anode materials	E_{corr} (V vs. Hg/HgO)	I_{corr} (A cm^{-2})	R_p (Ω cm^2)
Alloy	−1.751	9.841×10^{-3}	4.5
Pure Al	−1.364	2.695×10^{-2}	1.9

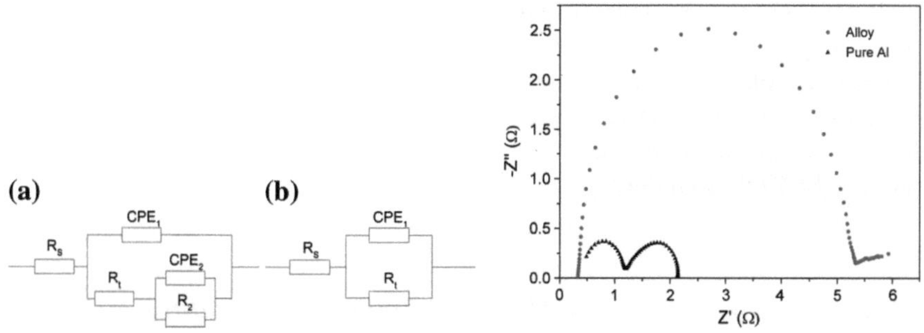

Fig. 3 EIS patterns of pure Al and alloy anodes in 4 M KOH solution (right-hand); Equivalent circuit of EIS of anodes for pure Al (**a**) and alloy (**b**) anodes (left-hand)

Table 3 EIS simulated values of anodes in 4 M KOH

	R_s (Ωcm^2)	CPE_1 (uFcm^{-2})	R_t (Ωcm^2)	CPE_2 (uFcm^{-2})	R_2 (Ωcm^2)	X^2
Pure Al	0.4227	15.62	0.7668	25,650	0.9812	3.09×10^{-4}
Alloy	0.3438	64.78	5.108	–	–	2.38×10^{-3}

polarization resistance: −1.751 V and 4.5 Ωcm^2, respectively. This means the alloy materials can serve as the anode to provide a large current to obtain a higher power density output than the pure aluminum.

Figure 3 shows EIS plots of the alloy and pure Al anodes obtained at open-circuit potential impedance measurements, as well as the equivalent circuits for simulating the process. Table 3 lists the simulation values obtained by ZSimpwin software. Here the capacitance element (*C*) which shows the double-layer capacitance is replaced by constant phase element (*CPE*) in order to have more precious fitting results. The χ^2 is the precision of the simulated data.

The EIS plots of pure Al anode consist of two loops: first one is the frequency loop that might result from the charge transfer reaction: Al $\rightarrow$ Al^{3+}. In the equivalent circuits it show as resistance (R_t) parallel to a constant phase element (CPE_1); the second one is the low frequency loop that might be due to the dissolution precipitation on the alloy anode surface. In this case, the equivalent component consists of a charge transfer resistance (R_2) that is parallel to a constant phase element (CPE_2), and R_s represents the resistance of KOH solution.

The EIS plots of alloy anode has only one high frequency loop. Compare with the pure Al, it does not have the low frequency loop which result from the dissolution precipitation on the alloy surface, which means the addition of the alloying element into the pure Al can increase the solubility of the corrosion film. In general, the higher R_t reflects a lower corrosion since the exchange current is directly associated with the electrochemical process of corrosion [12, 13]. It can be seen that the R_t of alloy anode is higher than the pure Al. This indicates that the corrosion rate of the alloy anode is slower than the pure Al, which is consistent with the potentiodynamic polarization curves test. Therefore, the addition of Mg, Ga, In, Sn elements can not only increase the electrochemical activity of anodes due to the dissolved redeposition mechanism of Ga, In [14], but also can reduce the corrosion rate in 4 M KOH solution.

Battery Performance with Heat Treatment Temperature

Figure 4 shows the discharge curves of the Al-air battery consist of as-cast alloy and pure Al anodes, respectively. It is clearly that the battery with the alloy anode can provide the higher discharge voltage, about 300 mV on average, at 20 $mAcm^{-2}$ with the additions of the Mg, Ga, In, Sn elements.

Figures 5, 6 and 7 show the discharge curves of the as-cast alloys which had been heat-treated at different temperatures and time periods. In general, the discharge voltage increases with increased heat-treatment temperatures, and this desired effect becomes stabilized with prolonged time from 6 to 12 h in heat treatment. Under this testing condition, the discharge voltage of the alloy heat-treated at 550 °C-12 h shows the highest value, about 1.62 V. In contrast, the as-cast alloy anode without any heat treatment gives lower and unstable discharge voltages after 2 h under the discharge experiments.

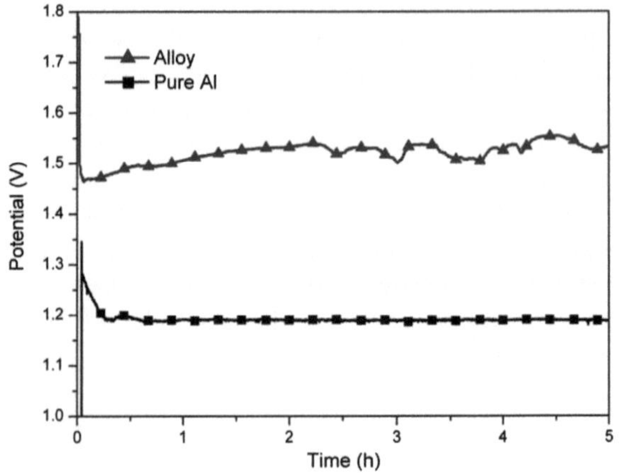

Fig. 4 Discharge curves of Al-air batteries with pure Al anode and as-cast alloy anode at 20 $mAcm^{-2}$

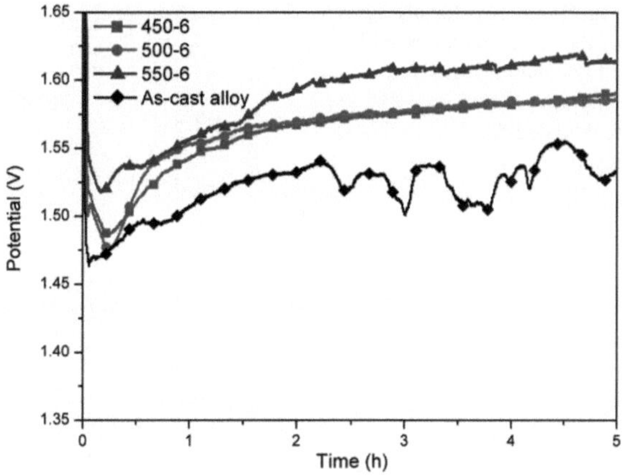

Fig. 5 Discharge curves of Al-air batteries with pure Al anode and alloy anode treated at different temperatures for 6 h at 20 $mAcm^{-2}$

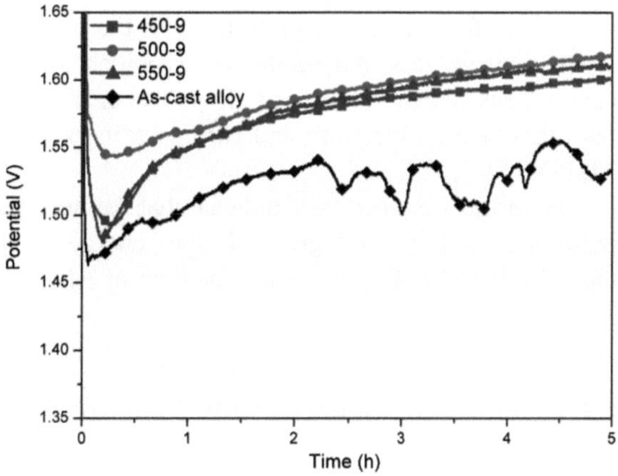

Fig. 6 Discharge curve of Al-air batteries with pure Al anode and alloy anode treated at different temperatures for 9 h at 20 $mAcm^{-2}$

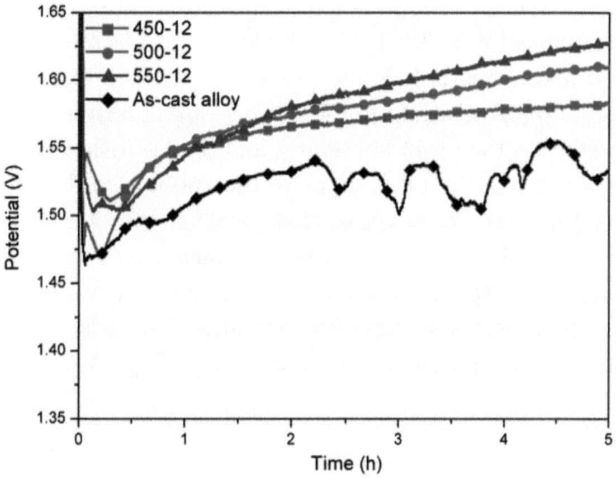

Fig. 7 Discharge curves of Al-air batteries with pure Al anode and alloy anode treated at different temperatures for 12 h at 20 $mAcm^{-2}$

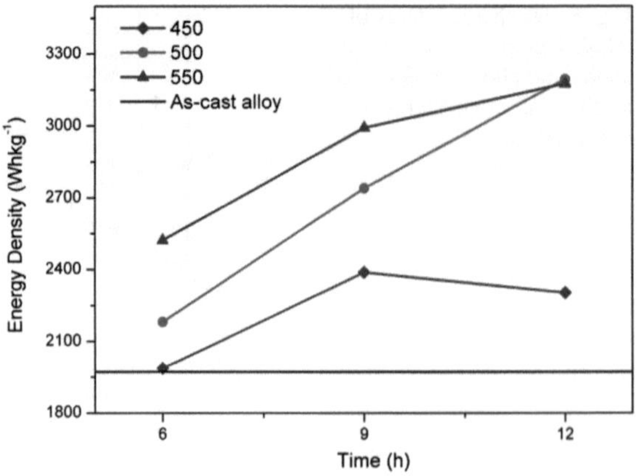

Fig. 8 Energy density curves of the Al-air batteries with alloys anodes treated at different heat treatments (calculated for 5 h discharge tests)

This fact is further approved, as shown in Fig. 8, by the energy density data (calculated for 5 h discharge tests) with the alloy anodes heat-treated for the different time periods. All of the alloys which have been heat-treated have higher energy density than the as-cast alloy. The energy density rises with the increase of both time and temperature and can reach its peak at 550 °C-12 h, which is about 3180 $Whkg^{-1}$.

The results obtained here indicate that the heat treatment has a positive effect on both the battery voltage and the energy density for the Al-0.2 Mg-0.2 Ga-0.4In-0.15Sn alloy as anode material in Al-air batteries.

Corrosion Morphology After Discharge Performance

Figure 9 shows the selected SEM micrographs on the surface morphologies of the different anodes after discharge at 20 $mAcm^{-2}$ current densities for 5 h. It can be seen that the morphology of the anode samples is associated with discharge performance. It is clearly that the alloy forms a reticular crack structure, while the pure Al anode has only some pits generated due to the corrosion process in 4 M KOH solution. According to the dissolved redeposition mechanism, Ga, In elements may gather in the grain boundary, and the corrosion at the grain boundary can be faster than that in the other areas of the anodes, so that the micro-cracks may get deeper and deeper. With those deep cracks, the charge transfer resistance crossing the reaction layer on the anode surface can reduce and hence the electrochemical activity of the alloy anode can increase due to the increased total reaction area. The surface morphologies observed after discharge performance are consistent with the trends of the potential-time curves in Figs. 5, 6, and 7.

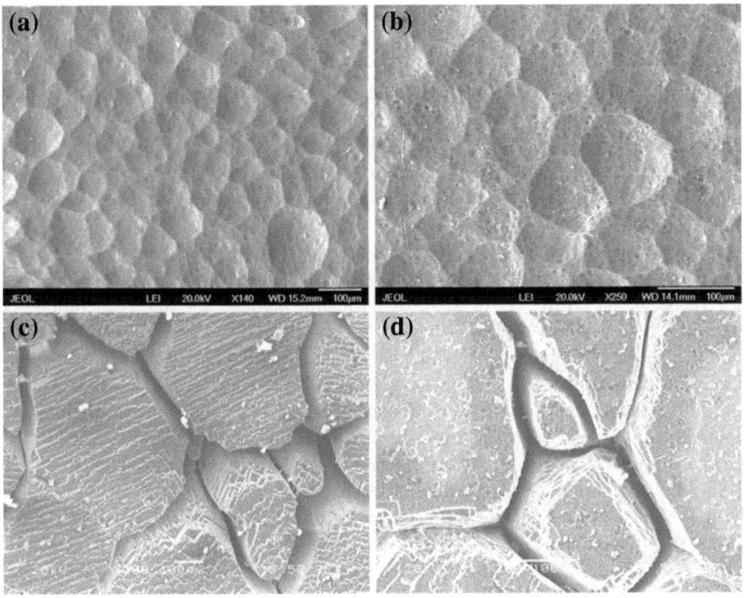

Fig. 9 Representative SEM micrographs of different anodes after discharge performance at 20 $mAcm^{-2}$ current densities for 5 h: **a** and **b** are pure Al; **c** and **d** are alloy heat-treated at 500 °C for 9 h

Conclusions

Al alloy of Al-0.2 Mg-0.2 Ga-0.4In-0.15Sn has been investigated as anode materials for Al-air batteries. The alloys as cast and heat-treated were characterized with OCP tests, electrochemical polarization, EIS tests and discharge performance, and the corrosion morphology after discharge performance was studied using SEM. The conclusions are as follows:

(1) The discharge voltage of as cast Al alloy anode containing 0.2 Mg-0.2 Ga-0.4In-0.15Sn is about 300 mV on average higher than that of the pure Al
(2) The discharge voltage and the energy density rise with increased heat treatment temperatures and prolonged time, which can be as high as about 1.62 V and about 3180 $Whkg^{-1}$, respectively, with heat treatment at 550 °C for 12 h.
(3) The microcracks due to the corrosion around the grain boundaries of the Al alloy anodes appear after the discharge performance in 4 M KOH solution, which may reduce the charge transfer resistance crossing the reaction layer on the anode surface.

Acknowledgements General support from aluminum industry in China and University of Science and Technology Beijing is acknowledged.

References

1. Egan DR, Ponce de leon, C, Wood RJK et al (2013) Developments in electrode materials and electrolytes for aluminium-air batteries. J Power Sources 236:293–310
2. Wang HZ, Leung DYC, Leung MKH (2012) Energy analysis of hydrogen and electricity production from aluminum-based processes. Appl Energy 90(1):100–105
3. Li QF, Bjerrum NJ (2002) Aluminum as anode for energy storage and conversion: a review. J Power Sources 110:1–10
4. Wang JB, Wang JM, Shao HB (2007) The corrosion and electrochemical behaviour of pure aluminium in alkaline methanol solutions. J Appl Electrochem 37(6):753–758
5. Kim H, Jeong G, Kim YU et al (2013) Metallic anodes for next generation secondary batteries. Chem Soc Rev 42:9011–9034
6. Abdel-Gaber AM, Khamis E, Abo-Eldahab H et al (2010) Novel package for inhibition of aluminium corrosion in alkaline solutions. Mater Chem Phys 124:773–779
7. Fan L, Lu H, Leng J et al (2015) Performance of Al-0.6Mg-0.05Ga-0.1Sn- 0.1In as Anode for Al-Air Battery in KOH Electrolytes. J Electrochem Soc 162(14):A2623–A2637
8. Sun Z, Lu H, Fan L et al (2015) Performance of Al-Air Batteries Based on Al-Ga, Al-In and Al-Sn Alloy Electrodes. J Electrochem Soc 162(10):A2116–A2122
9. Ma J, Wen J, Gao J et al (2014) Performance of Al-1Mg-1Zn-0.1Ga-0.1Sn as anodes for Air-battery. Electrochem Acta 129:69–75
10. Gundersen JTB, Aytac A, Nordlien JH et al (2004) Effect of heat treatment on electrochemical behaviour of binary aluminium mode alloys. Corros Sci 46:697–714
11. Afseth A, Nordlien JH, Scamans, et al (2002) Effect of heat treatment on electrochemical behavior of aluminium alloy AA3005
12. Hong T, Sun YH, Jepson WP (2002) Study on corrosion inhibitor in large pipelines under multiphase flow using EIS. Corros Sci 44:101–112
13. Osorio WR, Freitas ES, Garcia A (2013) EIS parameters and cell spacings of an Al-Bi alloy in NaCl solution. Electrochim Acta 108:781–787
14. Reboul MC, Gimenez PH, Rameau IJ (1984) A proposed activation mechanism for Al anodes. Corros 40(2):366–377

Part V
Extractive and Recovery Processing

A Current Efficiency Prediction Model Based on Electrode Kinetics for Iron and Copper During Copper Electrowinning

Zongliang Zhang, Joshua Werner and Michael Free

Abstract Copper electrowinning is an important recovery method in the copper industry, which accounts for a growing proportion of the world copper production. In copper electrowinning, not all the current is used for the deposition of Cu, in large part because of the existence of dissolved iron in the electrolyte. The reduction of ferric ions is often the main factor that causes the current efficiency to decrease. The current efficiency is influenced by parameters such as electrolyte temperature, current density, iron concentration, copper concentration, etc. To obtain optimal operating conditions, thus minimizing energy consumption, a copper electrowinning current efficiency model was established using empirical formulas and electrode kinetics. Electrochemistry, gas-liquid flow, transport phenomena and comparisons to experimental data are considered to enhance the accuracy of the model. The model can predict the current efficiency under a variety of conditions to provide valuable guidance for the optimization of process parameters.

Keywords Cu electrowinning · Modeling · Current efficiency
Fe kinetics

Introduction

Copper is a very important metal with wide application in modern society. In the industrial world, copper is produced with pyrometallurgical processes and hydrometallurgical processes. Hydrometallurgical processing is gaining increased

Z. Zhang (✉) · M. Free
University of Utah, Salt Lake City, USA
e-mail: zongliang.zhang@utah.edu

M. Free
e-mail: michael.free@utah.edu

J. Werner
University of Kentucky, Lexington, USA
e-mail: Joshua.werner@uky.edu

G. Lambotte et al. (eds.), *Materials Processing Fundamentals 2018*,
The Minerals, Metals & Materials Series,
https://doi.org/10.1007/978-3-319-72131-6_10

attention because of its high ability to economically process low-grade ores. Copper electrowinning is a significant link in copper hydrometallurgical process. Electrowinning is by nature complex and consumes a large amount of energy during Cu production. In modern Cu electrowinning processing, the Cu^{2+} ions in electrolyte typically deposit on stainless steel cathodes as metallic copper while oxygen is generated on Pb/Pb alloy anodes. Meanwhile, due to the existence of other impurities in electrolyte and short circuiting, the current efficiency is not 100%. The parasitic reaction of ferric ion reduction on the cathode is the most significant cause of current efficiency decreases in well run plants. The main reactions occurring on the cathodes and anodes in copper electrowinning cells are shown in Eqs. (1)–(4).

Cathode:

$$Cu^{2+} + 2e^- = Cu \tag{1}$$

$$Fe^{3+} + e^- = Fe^{2+} \tag{2}$$

Anode:

$$2H_2O = 4H^+ + O_2 + 4e^- \tag{3}$$

$$Fe^{2+} = Fe^{3+} + e^- \tag{4}$$

Generally, the current efficiency could be represented by Eq. (5).

$$\beta = \frac{i_{dep}}{i_{tot}} \tag{5}$$

where i_{dep} and i_{tot} are the expected metal deposition current density and total current density, respectively.

It can be challenging to evaluate the effects of parametric variations experimentally in production environments. Modeling and simulation have become a favorable alternative to physical experimentations. There have been a number of significant works published in this area. Ziegler and Evans [1–3] used RANS κ–l type Euler–Euler fluid flow to model the fluid flow in electrowinning system. Turbulent diffusivity is used to simulate the movement of bubbles in this work. Kisdnasamy and Neelakantaswamy [4] studied the effect of adhering gas bubbles on electrode surfaces on electrolyte conductivity. By comparing the results to Maxwell's and Bruggeman's equations, the experimental data shows general agreement with the overall trend. Nguyen et al. [5] present an electrowinning model considering time dependent diffusion and migration between parallel plates in a copper chloride system. The Nernst–Planck equation in a dimensionless form is used and the forced convection due to inlet and outlet flows was considered. Smyrl and Newman [6] investigated the influence of electrode kinetics on edges with a rotating disc electrode. This work is very helpful to determine when and how electrode kinetics should be used over primary current simplifications. Bark and Alavyoon [7] examined the problem in greater depth including transport via the Nernst–Planck equation and the addition of anodic and cathodic reaction kinetics as part of their model. However, the effects from bubbles was neglected. Marshall and Mocskos [8] produced a modeling electrochemical deposition with a simplified Navier–Stokes equation. The dendrite growth on a projecting filament was simulated. Dahlkild [9] provides additional insights by utilizing a two-phase equation of a Navier-Stokes type of model that accounts for gas fraction and gas diffusion. The model result appears to correlate well with the experimental data of Hine and Murakami [10]. Wuthrich et al. [11] consider the effect of bubbles on the voltage characteristics of the cell, presenting a quantitative model of the gas effects on the limiting current. Free et al. [12] discuss electrowinning modeling and performance, focusing on specific power consumption of deposition and current efficiency. Parasitic reactions were also considered with a model compared to experimental data. The work of Volgin and Davydov [13] presents significant illumination into the nature of the effects of speciation, which is insightful to the speciation and equilibrium reactions in electrolyte. Aldas et al. [14] studied a hydrolysis cell utilizing two phase flow and diffusion modifications due to gas fraction as a means of modifying conductivity with the Bruggman relation. The Butler–Volmer equation is also modified via the gas fraction to influence the electrode kinetics. Kawai et al. [15] modeled electrorefining based on transport flow with conductance and a simple speciation method. Leahy and Schwarz [16] provides an excellent comparison of an electrowinning model to previous work. Kim et al. [17] present a simplistic Navier-Stokes model of forced convection between two plates. Both electrorefining and electrowinning are considered. Hemmati et al. [18] studied multi-electrode electrorefining using a RANS (Reynolds averaged Navier-Stokes) k–ω turbulent mode. Leahy and Schwarz [19] in an expansion of his previous paper utilizes a RANS–SST (Shear Stress Transport) for time averaged turbulent flow. A copper sulfate electrowinning system is studied by Shukla and Free [20, 21]. Failure analysis statistics were utilized to predict time to short. The work of Robinson and Free [22] is included for the insights into deposit modeling. An expression describes the deposit thickness was proposed. The work of Zeng et al. [23, 24] focuses on modeling electrorefining and is primarily concerned with describing the transport of anode slimes to the cathode. Similar electrode kinetics and parameters will be used when developing current model. Najminoori et al. [25] uses a model with more than one electrode set that is very similar to that of Hemmati et al. [18] with RANS Euler–Euler k-ω as the computational fluid dynamic (CFD) component and the same type of transport equation. Hreiz et al. [26] present a model with a Euler–Lagrange model validated with PIV (Particle image velocimetry). Gas fractions and cell velocities were studied and captured. The numerical results were in general agreement with the experimentation.

In this paper, a comprehensive Cu electrowinning model was developed. It incorporated the electrochemistry, fluid flow, mass transfer, and bubble effects. Especially, the calculation of current efficiency was included in the model with two different methods. This has been seldom seen in previous models. With appropriate validation, this model could be used for the prediction of the current efficiency of Cu electrowinning.

Model Description

Cu electrowinning process is by nature complex. It is a combination of various processes: Electrochemical processes, mass transport, and two-phase flow, and exhibit a high degree of interaction. For example, the mass transport is affected by the fluid flow, and the density gradient caused by ion diffusion will affect the fluid flow. These processes can be described with several governing equations and boundary conditions in a CFD model that is solved with COMSOL Multiphysics package or similar software packages. A detailed description of the model has been discussed elsewhere [27]. Here the general structure of the model and the calculation of current efficiency are presented.

Governing Equations

Mass Transfer

In Cu electrowinning process, the mass transfer is comprised of three processes: diffusion, convection, and migration. In electrolyte, the governing equation for mass transfer in solution is the Nernst-Plank equation, as shown in Eq. (6)

$$N_i = -z_i u_i F C_i \nabla \Phi_l - D_i \nabla C_i + C_i v \tag{6}$$

where N_i, z_i, u_i, C_i, D_i are the flux density, charge, mobility, concentration, and diffusivity of species i, F is Faraday's constant, $\nabla \Phi_l$ is an electric field, ∇C_i is a concentration gradient, and v is the velocity vector.

Because there are no homogeneous reactions in the electrolyte, the material balance is governed by Eq. (7).

$$\frac{\partial C_i}{\partial t} + \nabla N_i = 0 \tag{7}$$

In the electrolyte, the current density is governed by Eq. (8).

$$i_{electrolyte} = -F^2 \nabla \Phi_l \sum z_i^2 u_i C_i - F \sum z_i D_i \nabla C_i + F v \sum z_i C_i \tag{8}$$

where $i_{electrolyte}$ is the current density in the electrolyte, and other variables are defined previously. Due to the electroneutrality of the electrolytic solution, the last term on the right is zero ($\sum z_i C_i = 0$). Therefore,

$$i_{electrolyte} = -F^2 \nabla \Phi_l \sum z_i^2 u_i C_i - F \sum z_i D_i \nabla C_i \tag{9}$$

The species considered in the model are: $CuHSO_4^+$, SO_4^{2+}, HSO_4^-, H^+, $FeSO_4^+$, and Fe^{2+}.

The diffusion coefficients for copper and iron were calculated using the work of Moats et al. [28]. The work of Moats et al. was reviewed and regression was utilized to develop Eq. (10).

$$\begin{aligned} \log D_{0,\mathrm{Cu}} = & -0.676 - 0.481 \log(C_{\mathrm{H_2SO_4}} Mw_{\mathrm{H_2SO_4}}) \\ & - 0.156 \log(C_{\mathrm{Cu}} Mw_{\mathrm{Cu}}) + 0.9885 \left(\frac{-8340.61}{8.314T} \right) \end{aligned} \tag{10}$$

where $D_{0,\mathrm{Cu}}$ is in cm^2/s, $C_{\mathrm{H_2SO_4}}$ and C_{Cu} are the initial concentrations in mol/m^3, $Mw_{\mathrm{H_2SO_4}}$ and Mw_{Cu} are in kg/mol, T is temperature in K. This equation has the following range: $C_{\mathrm{Cu}} = 35 - 60\,(\mathrm{g/L})$, $C_{i,\mathrm{H_2SO_4}} = 160 - 250\,(\mathrm{g/L})$, T = 40–65 °C. This expression demonstrates a reasonable fit of the data.

For iron diffusion coefficients,

$$D_{\mathrm{Fe^{3+}}} = 10^{-6}(-3.01 + 1.58C_{\mathrm{Fe}} - 0.0015C_{\mathrm{Cu}} + 0.0082C_{\mathrm{H_2SO_4}} + 0.142T) \tag{11}$$

where $D_{\mathrm{Fe^{3+}}}$ is in cm^2/s, C_{Fe}, C_{Cu} and $C_{\mathrm{H_2SO_4}}$ are the initial concentrations in g/L, T is temperature in °C.

$$D_{\mathrm{Fe^{2+}}} = 1.18 D_{\mathrm{Fe^{3+}}} \tag{12}$$

For other ions, the diffusion coefficients were calculated by solving for the ionic radius using the Stokes-Einstein relation.

The effect of diffusion is modified because of the presence of oxygen bubbles formed at the anode. For this work the approach is to make diffusion modification a function of gas fraction in the electrolyte. Table 1 shows several equations for this purpose. These were adapted from the work of Hammoudi et al. [29] which originally represented conductivity as a function of gas fraction. The selection of any of these equations at gas fractions <0.1 is rather arbitrary as there is little variation. In this work, the Maxwell equation from Table 1 was utilized.

Table 1 Diffusion modification based on gas fraction [29]

Relative diffusion	Author
$D = D_0 \frac{(1-\varepsilon)}{(1+\varepsilon)}$	Rayleigh
$D = D_0 \frac{(1-\varepsilon)}{(1+0.5\varepsilon)}$	Maxwell
$D = D_0 \frac{8*(1-\varepsilon)*(2-\varepsilon)}{(4+\varepsilon)(4-\varepsilon)}$	Tobias
$D = D_0(1-\varepsilon)^{1.5}$	Bruggeman
$D = D_0(1 - 1.5\varepsilon + 0.5\varepsilon^2)$	Prager

Migration was modeled by relating mobility to diffusion using the Nernst-Einstein relation, as shown in Eq. (13).

$$u = \frac{zFD}{RT} \tag{13}$$

Fluid Flow

The fluid flow in Cu electrowinning cells is often affected by three factors: inlet and outlet flow, bubbles, and electrolyte density gradient. The form of the bubbly flow equations used assumes that the fluid is non-compressible and utilizes the following additional assumptions [30]:

- The gas phase density is insignificant compared to the liquid phase density
- The motion between the phases is determined by a balance of viscous and pressure forces
- Both phases are in the same pressure field
- Low gas concentration
- No turbulence modeling.

The resulting equation is shown in Eq. (14) [31]:

$$\phi_l \rho_l \frac{\partial u_l}{\partial t} + \phi_l \rho_l (u_l \cdot \nabla) u_l = \nabla \cdot \left[-p2I + \phi_l \rho_l \left(\nabla u_l + (\nabla u_l)^T \right) \right] + \phi_l \rho_l g + F \tag{14}$$

The continuity equation simplified via the low gas concentration assumption is:

$$\phi_l \nabla \cdot u_l = 0 \tag{15}$$

where u_l is the velocity vector (SI unit: (m/s), p is the pressure (SI unit: Pa), φ_l is the phase volume fraction (SI unit: m^3/m^3), ρ_l is the density (SI unit: kg/m^3), g is the gravity vector (SI unit: m/s^2), F is any additional volume force (SI unit: N/m^3), I is the identity matrix.

Geometry

For simplification and acceleration of the calculation process, a 2D test cell geometry with only one set of anode and cathode was used, as shown in Fig. 1. The cathode is 0.17 m long and the anode is 0.165 m long. Electrode gap is 2.54 cm. Because of the limitation of 2D geometry, the inlet and outlet are set at the top and bottom respectively. This arrangement has little effect on the fluid flow because the inlet and outlet flow rates are relatively small compared with the bubbly flow.

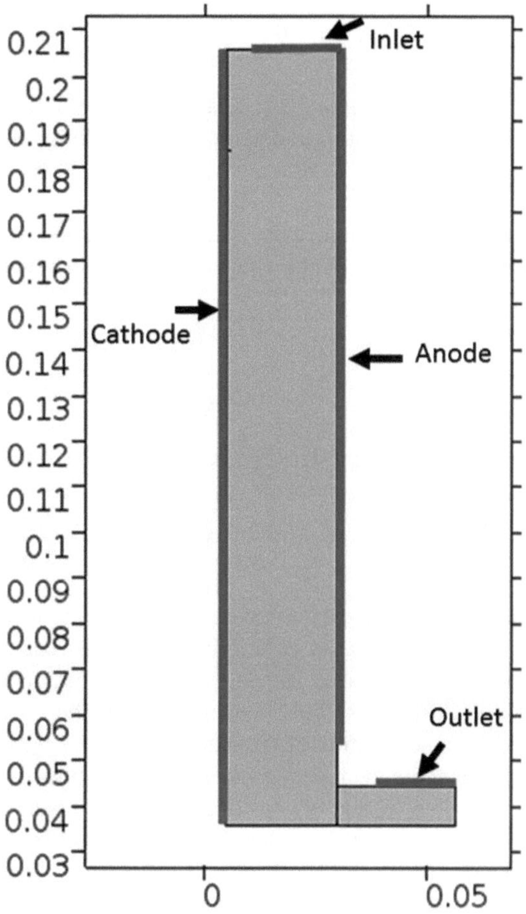

Fig. 1 Geometry of the electrowinning cell in the model (Units in m)

Boundary Conditions for Mass and Momentum Transfer

Electrodes

Equilibrium Potential

In order to determine cell voltages, the equilibrium potential of both the anodic and cathodic reactions needed to be calculated. The equilibrium potential or E is given by the Nernst equation as Eq. (16).

$$E = E^0 - \frac{2.303RT}{nF}\log\left(\frac{a_{products}}{a_{reactants}}\right) \quad (16)$$

This equation uses the standard thermodynamic potential along with the product and reactant activities to determine the equilibrium potential for a half-cell reaction.

Cathode Kinetics

For simplification, it is assumed that copper and iron reduction occur on the cathode according to the chemical reactions shown in Eqs. (17) and (18).

$$\mathrm{CuHSO}^+_{4(aq)} + 2\mathrm{e}^- = \mathrm{Cu} + \mathrm{HSO}_4^- \quad (17)$$

$$\mathrm{FeSO}_4^+ + \mathrm{e}^- = \mathrm{Fe}^{2+} + \mathrm{SO}_4^{2-} \quad (18)$$

The local current density is calculated with an adapted Butler-Volmer Eq. (19).

$$i_{loc} = i_0\left[\frac{C_{R,S}}{C_{R,B}}\exp\left(\frac{\alpha_a zF}{RT}\eta\right) - \frac{C_{O,S}}{C_{O,B}}\exp\left(\frac{\alpha_c zF}{RT}\eta\right)\right] \quad (19)$$

where i_{loc} is the local current density at the interface (also called the charge transfer current density), i_0 is equilibrium exchange current density, $C_{R,S}$ is the surface concentration reduced species, $C_{R,B}$ is the bulk concentration of the reduced species, $C_{O,S}$ is the surface concentration of the oxidized species, $C_{O,B}$ is the bulk oxidant concentration of the oxidized species, α_a is the anodic symmetry factor, α_c is the cathodic symmetry factor, z is the number of electrons transferred in the rate limiting step (nearly always one), F is the Faraday constant, R is the gas constant, T is the absolute temperature, and η is the overpotential.

In reviewing this work in light of Newman's equation for the copper reaction [32] it was determined that appropriate cathodic and anodic charge transfer coefficients are 0.545 and 1.455, respectively. For the exchange current density, Eq. (20) was used [33]:

$$i_{of} = i_{oi}\left(\frac{a_{Cuf}}{a_{Cui}}\right)^{1-\frac{\beta}{2}} \quad (20)$$

where, i_{of} is the exchange current density at the desired concentration, i_{oi} is the exchange current density at the reference concentration, a_{Cuf} is the final activity, a_{Cui} is the reference activity at i_{oi}, β is the symmetry factor. From the works cited above i_{oi} is set at 1 mol/l was 100 A/m^2. Exchange current density of iron are adopted from Cifuentes's paper [34].

Anode Kinetics

For the anodic reaction, the work of Laitinen [35] was referenced. This work represents experimental determination of i_0 at different temperatures on a Pb–PbO_2 electrode. Equation (21) was developed based on the data from this literature.

$$i_0 = 7 \times 10^{-9} T - 2 \times 10^{-6} \tag{21}$$

where, i_0 is the exchange current density in A/m^2, T is temperature in K.

Current Efficiency

Two different methods were used to calculate the current efficiency of Cu electrowinning process. One is a comparative expression of the diffusion of ferric to copper ions, the other involves the calculation of iron species transport and reaction kinetics. The comparative equation was adopted from the work of Ettel et al. [36]. This method provides an easy approach to determine current efficiency without direct calculation of Fe^{3+} transport and consumption with Butler-Volmer kinetics. The equations for current efficiency in these two methods are shown in Eqs. (22) and (23).

$$\beta = 1 - \frac{FC_{bulk,\mathrm{Fe}^{3+}} \left(\frac{N_{\mathrm{Cu}2+}}{C_{bulk,\mathrm{Cu}2+} - C_{electrode,\mathrm{Cu}2+}} \right) \left(\frac{D_{\mathrm{Cu}2+}}{D_{\mathrm{Fe}3+}} \right)^{\frac{3}{4}}}{i_{total}} \tag{22}$$

where, β is current efficiency, F is the Faraday constant, $C_{bulk,\mathrm{Fe}^{3+}}$ and $C_{bulk,\mathrm{Cu}^{2+}}$ are the bulk concentrations of Fe and Cu respectively, $C_{electrode,\mathrm{Cu}^{2+}}$ is the Cu concentration at electrode surface, $N_{\mathrm{Cu}^{2+}}$ is the Cu flux, $D_{\mathrm{Cu}^{2+}}$ and $D_{\mathrm{Fe}^{3+}}$ are the diffusion coefficients of Cu and Fe respectively, i_{total} is the total current density.

$$\beta = 1 - \frac{i_{\mathrm{Fe}}}{i_{total}} \tag{23}$$

where, i_{Fe} is the iron reaction current density.

Inlet, Outlet and Walls

Liquid and gas phases have different inlets and outlets. For the liquid phase, inlet and outlet are shown in the model geometry shown in Fig. 1. For the gas phase, the inlet is anode, while outlet is the whole top line and the liquid outlet. Other boundaries are set as no-slip boundary walls.

Operating Conditions

The operating conditions used in the model is the same as what it used in the test cell, as shown in Table 2.

Results and Discussion

General Results and Validation

Detailed results of this kind of model have been shown and discussed by Werner [27]. The focus of this paper is on the prediction of current efficiency. Figure 2 shows the electrolyte velocity in the cell with the pronounced vortex at the top of the cathode. This vortex enhances the local mass transport conditions which leads to the formation of a copper ridge. This phenomenon has been verified by the measurements of experimental obtained copper deposits [27].

Figure 3 shows the gas fraction in the cell with the majority exiting the electrolyte near the anode and at the top of the cell. This figure shows that there exists only a small region where the gas concentration exists in any appreciable concentration. That region is next to the anode. As such it is felt that the low gas concentration assumption remains valid for this model.

Cu concentration distribution along cathode surface is shown in Fig. 4. Because of the Cu deposition on cathode, the surface concentration is lower than bulk or inlet concentration. However, at the cathode top, Cu concentration is higher than other locations because of the enhanced mass transport caused by the vortex.

Figure 5 shows the current density as a function of position comparing the model results to those of the experiment. The model and experimental current distribution show similar distribution and magnitude at specific position.

Table 2 Cell operation conditions

Cathode height	0.17 m
Anode height	0.165 m
Temperature	40 °C
Current density	277.7 A/m^2
Copper concentration	40 g/L
Sulfuric acid concentration	180 g/L
Inlet flow rate	29.5 mL/min
Distance between anode and cathode	2.54 cm

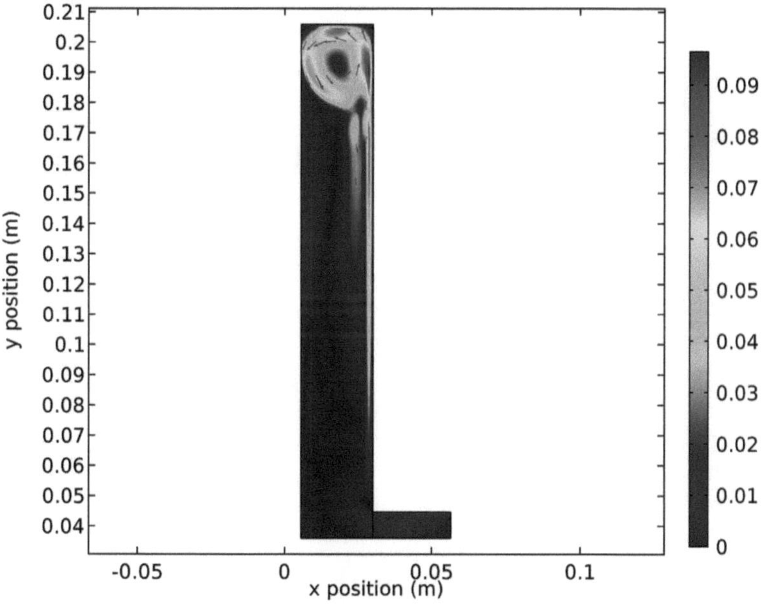

Fig. 2 Modeled cathode cell electrolyte velocity (m/s) with arrows as velocity vectors

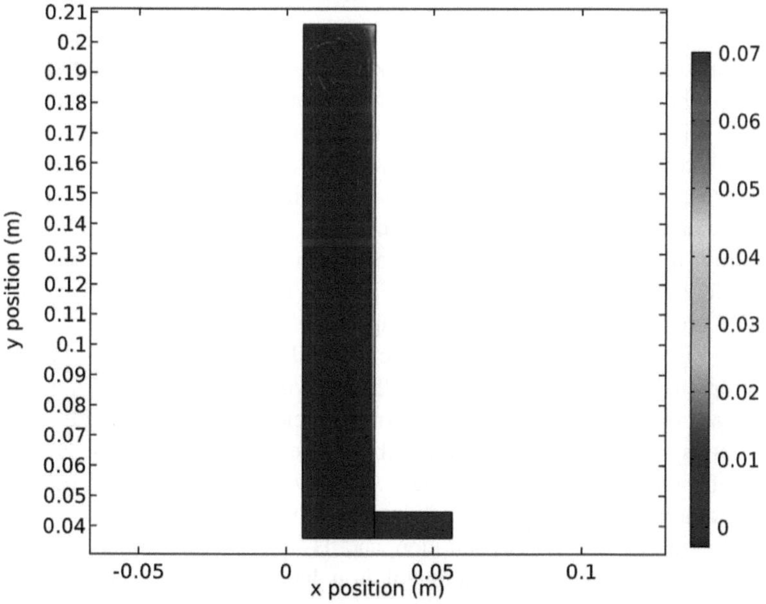

Fig. 3 Modeled cathode cell gas fraction with arrows as velocity vectors

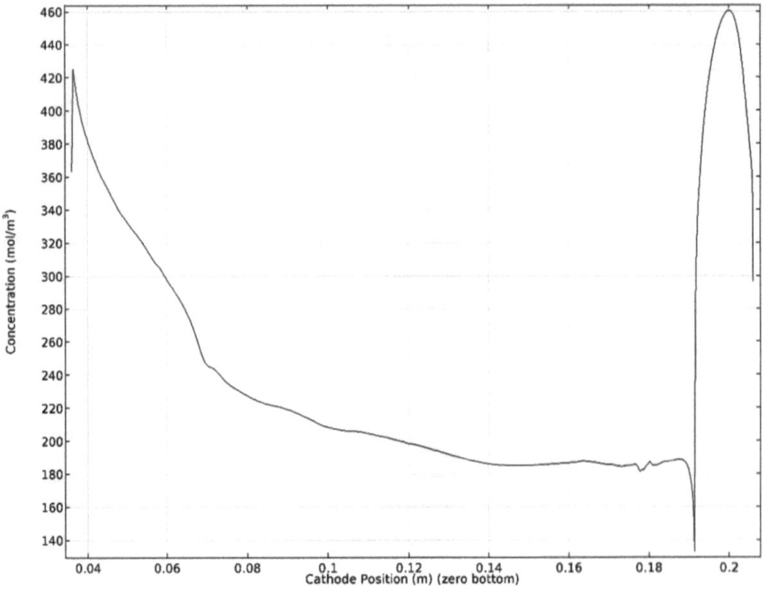

Fig. 4 Cu concentration distribution along cathode

Current Efficiency Prediction with a Comparative Method

In the comparative method model, Fe^{2+} and Fe^{3+} were not included in the electrode kinetics of the model, while the calculation of current efficiency was realized with Eq. (22). With this model, various operating parameters sets, i.e. iron concentration, copper concentration, temperature, and sulfuric acid concentration, were used to predict the current efficiency under certain conditions. The predicted current efficiency under various condition combinations are shown and discussed below.

Figure 6 illustrates the current efficiency with three different temperatures and three different copper concentrations. Figure 6 also shows the effects of these variables on current efficiency. As temperature goes up, the current efficiency decreases significantly. This could be explained by the change in the cathodic concentration gradient and diffusion coefficient with temperature in Eq. (22). With the increase of copper concentration, current efficiency decreased a little bit. From these results, it is apparent that the temperature has a much stronger effect on current efficiency than Cu concentration. However, this may not be true because results are derived from the relationship shown in in Eq. (22).

The influences of Fe concentration and H_2SO_4 concentration are shown in Fig. 7. Obviously, the concentration of iron has the most important effects on current efficiency because the parasitic reaction of iron on cathode. The comparative equation predicted a current efficiency drop from about 97.5% to about 90% while the iron concentration goes from 1 to 3 g/L. However, the predicted influence of

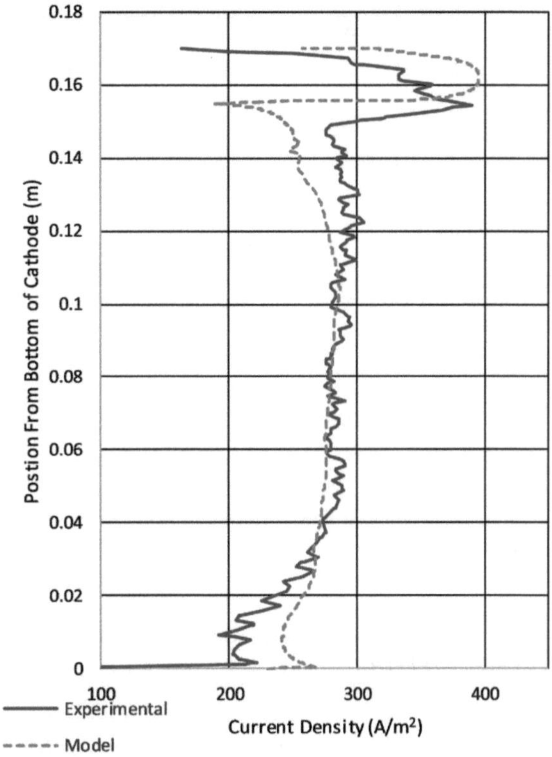

Fig. 5 Average cathodic current density (60–90 s) comparison model to experimental data [27]

iron is a little bit higher than experimental values. The concentration of H_2SO_4 shows no apparent effects on current efficiency.

Figures 8 and 9 show the other two different combinations of the parameters discussed above. Similar trend in Figs. 6 and 7 can be observed in these figures. With the above discussion, it is reasonable to state that the comparative equation model could provide a reasonable prediction of the current efficiency in Cu electrowinning processes under various combination of operating parameters. However, due to its limitations, it may not be a very good current efficiency prediction method over a wide range of different kinds of operating parameters. Caution should be observed with the usage of such an equation.

Current Efficiency Prediction with Kinetics Method

In contrast to the comparative equation model, a direct kinetics based model predicts the current efficiency from a fundamental level because it calculates the iron

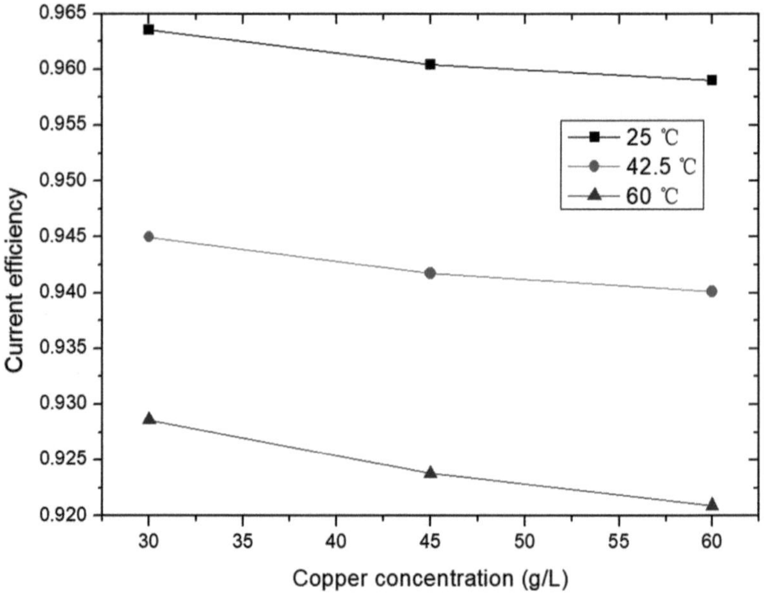

Fig. 6 Current efficiency change with Cu concentration under different temperatures (C_{Fe} = 2 g/L, $C_{H_2SO_4}$ = 190 g/L)

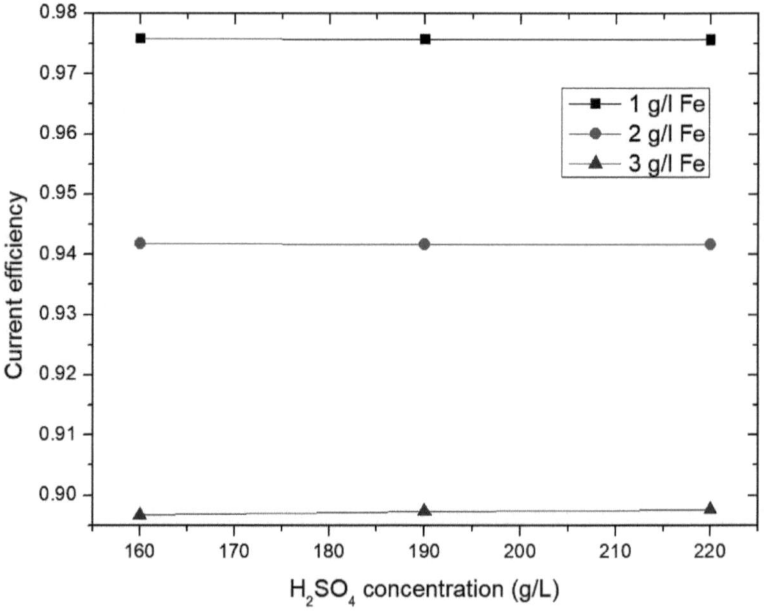

Fig. 7 Current efficiency change with H_2SO_4 concentration under different Fe concentrations (C_{Cu} = 45 g/L, T = 42.5 °C)

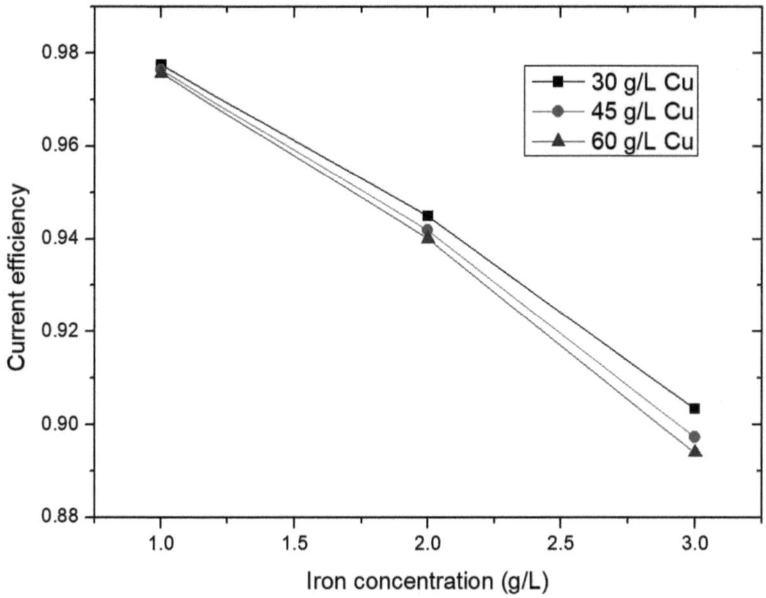

Fig. 8 Current efficiency change with Fe concentration under different Cu concentrations ($C_{H_2SO_4} = 190$ g/L , T = 42.5 °C)

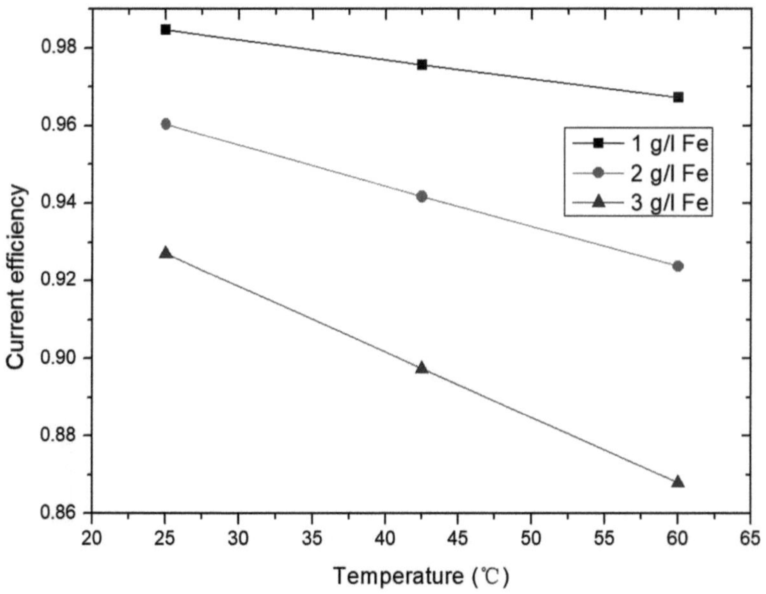

Fig. 9 Current efficiency change with temperature under different Fe concentrations ($C_{H_2SO_4} = 190$ g/L , $C_{Cu} = 45$ g/L)

electrode reaction rates based on the underlying thermodynamics, mass transport, and kinetics, which is the same process for the copper reaction calculation. Figures 10 and 11 show the Fe^{3+} and Fe^{2+} concentration distribution respectively in the electrowinning cell when the iron concentration is 6 g/L. As we can see, in the bulk electrolyte, the concentration is comparatively uniform except for some disturbance at the top of the cell due to the vortex. It is clear that the Fe^{3+} consumed and transformed to Fe^{2+} at cathode surface. Accordingly, Fe^{2+} is depleted at anode surface and transformed to Fe^{3+}. There are concentration boundary layers on both cathode and anode. The concentration of Fe^{3+} is close to 0 at cathode surface, and the concentration of Fe^{2+} is close to 0 at anode surface, indicating that the iron reaction on both cathode and anode are running at its limiting current density.

Similar with the comparative equation model, the current efficiencies under different operating conditions were predicted. Because of the intensive requirement for computational power, only the influence of the most important parameter iron concentration was studied. The studied iron concentration is 1, 2, 4, and 6 g/L. The detailed current and current efficiency distribution along the cathode is available in this model because the iron species are included as other species. The current density resulting from the consideration of the iron reaction along cathode is shown in Fig. 12. The higher the iron concentration, the more current was consumed in the

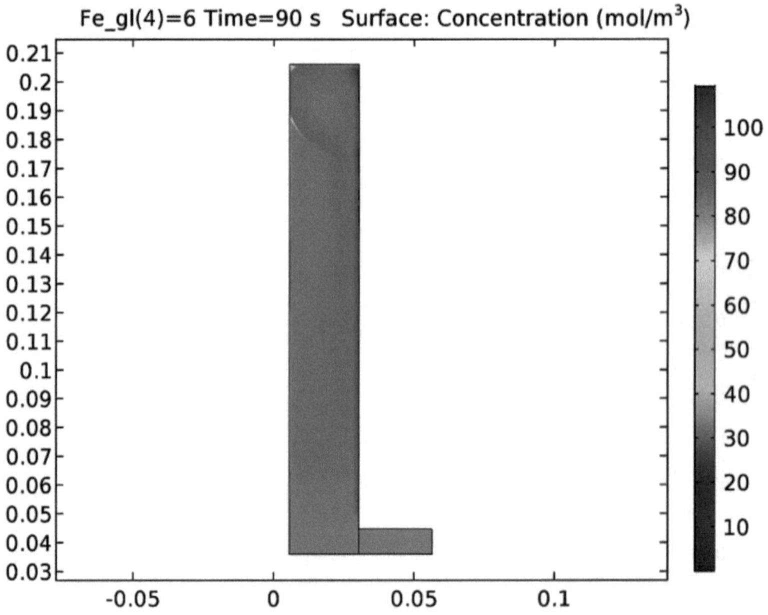

Fig. 10 Fe^{3+} concentration distribution in the electrowinning cell at pseudo-steady state

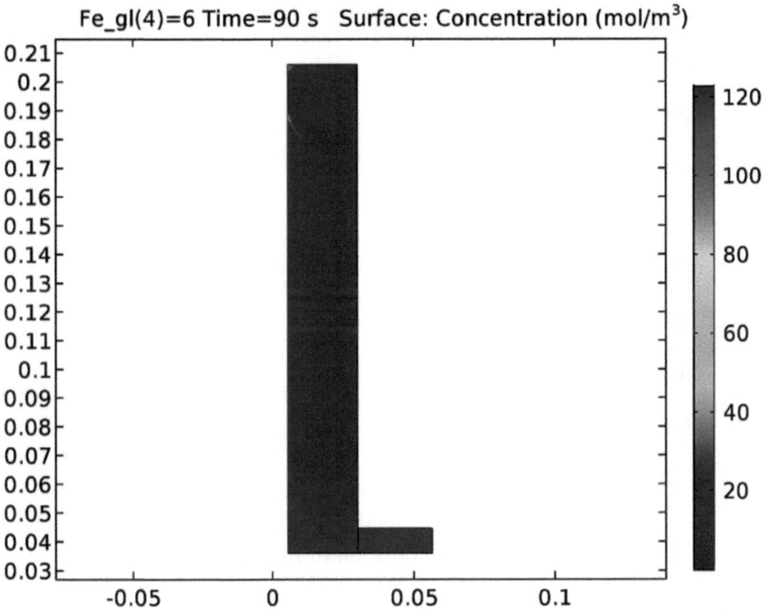

Fig. 11 Fe^{2+} concentration distribution in the electrowinning cell at pseudo-steady state

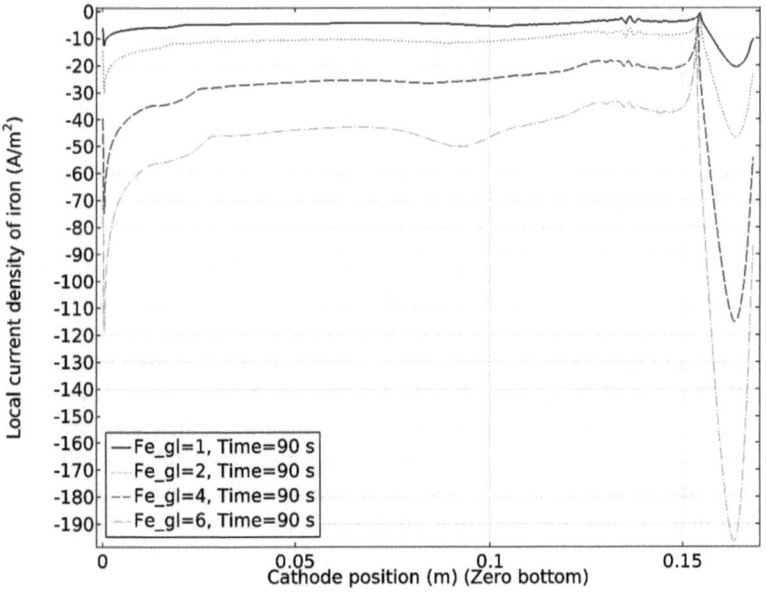

Fig. 12 The current density of iron reaction along cathode under different iron concentrations at pseudo-steady state

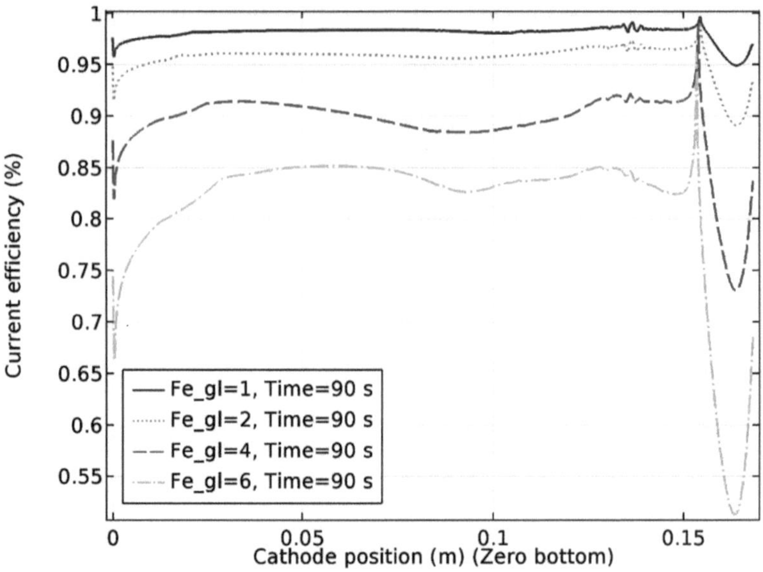

Fig. 13 The current efficiency of along cathode under different iron concentrations at pseudo-steady state

iron reaction. At the top of the cathode, because of the existence of vortex, and the enhanced transport of the iron species more current was used for iron reaction, lowering the current efficiency.

Based on the current density of the Cu deposition reaction and the iron reduction reaction, the local current efficiencies under different iron concentrations were calculated and shown in Fig. 13. In the middle part of the cathode, the current efficiency is uniform and relatively steady. Accordingly, at the top cathode, the current efficiency drops dramatically.

Das and Krishna [37] did a thorough inspection of the influence of iron concentration on Cu electrowinning current efficiency with a small test cell. The predicted current efficiencies from both models under different iron concentrations were compared with three different sets of current efficiencies from the experiments. For different sets of experimental data, the current efficiency varies even under the same iron concentration. Generally, the predicted current efficiencies are within the same range of the experimental data and show same trend with iron concentration. Therefore, the current efficiency models are deemed reliable in the studied range of data and may be used for operating prediction and optimization (Fig. 14).

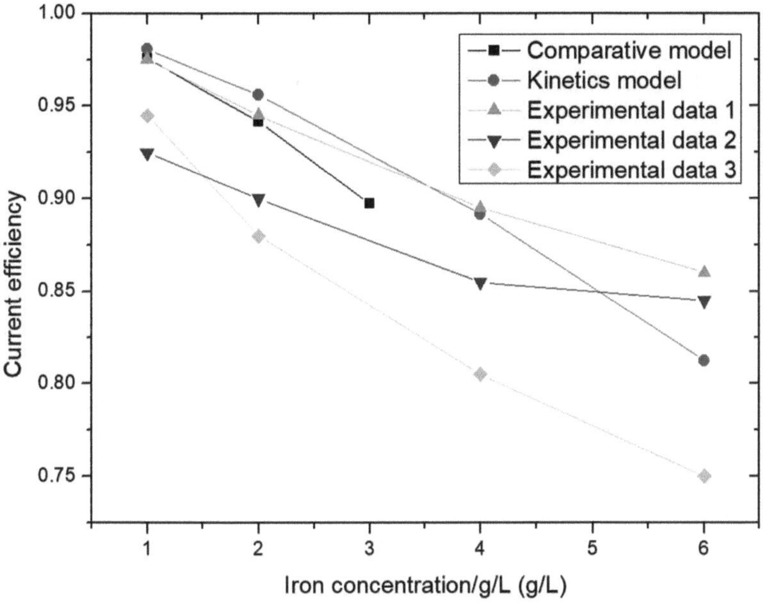

Fig. 14 Current efficiency comparison among model predicted data and experimental data [37]

Summary

This paper introduced a current efficiency model which incorporated the prediction of current efficiency utilizing two different methods. Electrochemistry, two phase flows, and mass transfer were modeled. Two methods of current efficiency prediction were evaluated and the relative advantages and limitations were discussed. The comparative equation model is simple and less computational intensive, and it gives a general prediction of the current efficiency under reasonable parametric ranges. However, its prediction may not be reliable under some conditions. The current efficiency model based on the iron species mass transfer, reaction thermodynamics, and kinetics is computational intensive, but it can provide more detailed results. This is because it treats iron the same way as copper or other ions. Whereas the comparative method infers the transport of iron from that of copper. In general, the predicted current efficiency shows a reasonable decreasing trend with iron concentration. For the completeness of the model, this promising current efficiency model could be developed to full size and used in the prediction of the current efficiency under new operating parameters. This is beneficial for performance prediction and process optimization.

Acknowledgements The authors wish to acknowledge the partial funding for this project from Amira.

References

1. Ziegler D, Evans JW (1986) Mathematical modeling of electrolyte circulation in cells with planar vertical electrodes I. Electrorefining cells. J Electrochem Soc 133(3):559–566
2. Ziegler D, Evans JW (1986) Mathematical modeling of electrolyte circulation in cells with planar vertical electrodes II. Electrowinning cells. J Electrochem Soc 133(3):567–576
3. Ziegler D (1984) A study of electrowinning and electrorefining cell hydrodynamics. PhD Thesis University of California, Berkeley
4. Kisdnasamy S, Neelakantaswamy PS (1984) A quantitative assessment of bubble-curtain effects upon interelectrode resistance of a conductimetric cell. J Appl Electrochem 14(6): 749–754
5. Nguyen, T. V., Walton, C. W., White, R. E., & Van Zee, J. (1986). Parallel-Plate Electrochemical Reactor Model A Method for Determining the Time-Dependent Behavior and the Effects of Axial Diffusion and Axial Migration. *Journal of the Electrochemical Society*, *133*(1), 81–87.
6. Smyrl WH, Newman J (1989) Current distribution at electrode edges at high current densities. J Electrochem Soc 136(1):132–139
7. Bark FH, Alavyoon F (1995) Free convection in an electrochemical system with nonlinear reaction kinetics. J Fluid Mech 290:1–28
8. Marshall G, Mocskos P (1997) Growth model for ramified electrochemical deposition in the presence of diffusion, migration, and electroconvection. Phys Rev E 55(1):549
9. Dahlkild AA (2001) Modelling the two-phase flow and current distribution along a vertical gas-evolving electrode. J Fluid Mech 428:249–272
10. Hine F, Murakami K (1980) Bubble effects on the solution IR drop in a vertical electrolyzer under free and forced convection. J Electrochem Soc 127(2):292–297
11. Wüthrich R, Comninellis C, Bleuler H (2005) Bubble evolution on vertical electrodes under extreme current densities. Electrochim Acta 50(25):5242–5246
12. Free ML, Bhide R, Rodchanarowan A, Phadke N (2006) Electrochemical modeling of electrowinning performance. In: 2006 TMS Fall Extraction and Processing Division: Sohn International Symposium
13. Volgin VM, Davydov AD (2008) The limiting current density of copper electrodeposition on vertical electrode under the conditions of electrolyte natural convection. Russ J Electrochem 44(4):459–469
14. Aldas K, Pehlivanoglu N, Mat MD (2008) Numerical and experimental investigation of two-phase flow in an electrochemical cell. Int J Hydrogen Energy 33(14):3668–3675
15. Kawai S, Fukunaka Y, Kida S (2010) Numerical calculation of transient current density distribution along vertical plane electrode in $CuSO_4$–H_2SO_4 electrolyte solution. J Electrochem Soc 157(3):F40–F48
16. Leahy MJ, Schwarz MP (2010) Experimental validation of a computational fluid dynamics model of copper electrowinning. Metall Mater Trans B 41(6):1247–1260
17. Kim KR, Choi SY, Paek S, Park JY, Hwang IS, Jung Y (2013) Electrochemical hydrodynamics modeling approach for a copper electrowinning cell. Int J Electrochem Sci 8(11):12333–12347
18. Hemmati H, Mohebbi A, Soltani A, Daneshpajouh S (2013) CFD modeling of the electrolyte flow in the copper electrorefining cell of Sarcheshmeh copper complex. Hydrometallurgy 139:54–63
19. Leahy MJ, Schwarz MP (2014) Flow and mass transfer modelling for copper electrowinning: development of instabilities along electrodes. Hydrometallurgy 147:41–53
20. Shukla A (2013) Modeling and measuring electrodeposition parameters near electrode surfaces to facilitate cell performance optimization. Master Thesis University of Utah.
21. Shukla A, Free ML (2014) Modeling and measuring electrodeposition parameters near electrode surfaces to facilitate cell performance optimization, In: Proceedings of 7th International Symposium on Hydrometallurgy, vol 2. MetSoc, Westmont, Canada, pp 67–68

22. Robison M, Free ML (2014) Modeling and experimental validation of electroplating deposit distributions from copper sulfate solutions. ECS Trans 61(21):27–36
23. Zeng W, Werner J, Free ML (2015) Experimental studies on impurity particle behavior in electrolyte and the associated distribution on the cathode in the process of copper electrorefining. Hydrometallurgy 156:232–238
24. Zeng W, Free ML, Werner J, Wang S (2015) Simulation and validation studies of impurity particle behavior in copper electrorefining. J Electrochem Soc 162(14):E338–E352
25. Najminoori M, Mohebbi A, Arabi BG, Daneshpajouh S (2015) CFD simulation of an industrial copper electrowinning cell. Hydrometallurgy 153:88–97
26. Hreiz R, Abdelouahed L, Fuenfschilling D, Lapicque F (2015) Electrogenerated bubbles induced convection in narrow vertical cells: PIV measurements and Euler-Lagrange CFD simulation. Chem Eng Sci 134:138–152
27. Werner JM (2017) Modeling and validation for optimization of electrowinning performance, PhD Thesis University of Utah
28. Moats MS, Hiskey JB, Collins DW (2000) The effect of copper, acid, and temperature on the diffusion coefficient of cupric ions in simulated electrorefining electrolytes. Hydrometallurgy 56(3):255–268
29. Hammoudi M, Henao C, Agbossou K, Dubé Y, Doumbia ML (2012) New multi-physics approach for modelling and design of alkaline electrolyzers. Int J Hydrogen Energy, 37 (19):13895–13913
30. CFD Module Users Guide, (2015) version 5.2, Comsol
31. Comsol multiphysics Qr v. 5.2a, www.comsol.com
32. Newman J, Thomas-Alyea KE (2012) Electrochemical systems. Wiley, New York
33. Mattsson E, Bockris JM (1959) Galvanostatic studies of the kinetics of deposition and dissolution in the copper+copper sulphate system. Trans Faraday Soc 55:1586–1601
34. Cifuentes L, Castro JM, Crisóstomo G, Casas JM, Simpson J (2007) Modelling a copper electrowinning cell based on reactive electrodialysis. Appl Math Model 31(7):1308–1320
35. Laitinen T, Pohl JP (1989) The kinetics of oxide growth and oxygen evolution during potentiostatic polarization of lead in H2SO4 electrolytes—investigations on the temperature dependence. Electrochim Acta 34(3):377–385
36. Ettel VA, Tilak BV, Gendron AS (1974) Measurement of cathode mass transfer coefficients in electrowinning cells. J Electrochem Soc 121(7):867–872
37. Das SC, Krishna PG (1996) Effect of Fe (III) during copper electrowinning at higher current density. Int J Miner Process 46(1–2):91–105

The K_2SO_4–$CaSO_4$ System and Its Role in Fouling and Slagging During High-Temperature Processes

Fiseha Tesfaye, Daniel Lindberg and Leena Hupa

Abstract Currently, there is an increasing interest to produce energy mainly from renewable sources such as biomass. However, fouling, slagging and corrosion threaten long-term operation availability and costs of biomass power plants. Alkali metal elements in the biomass fuel and the ash fusion behavior are the two major origins contributing to slagging during high-temperature biomass combustion. Accumulated slags decrease thermal efficiency of superheaters. These slags often constitute a considerable percentage of complex inorganic phases such as $K_2Ca_2(SO_4)_3$. However, thermodynamic properties of these inorganic phases and their combined effect, which would help to solve the fouling, slagging and high-temperature corrosion related problems in biomass combustion processes, are not well known. In the present work, thermodynamics and phase equilibria of selected phases in the K_2SO_4–$CaSO_4$ system were both critically reviewed and experimentally studied. The obtained results are presented and discussed.

Keywords Sulfates · Thermodynamic properties · Renewable energy

Introduction

Today, several countries aim to obtain considerable percentage of their energy consumptions from renewable sources. Aside from the usage of solar, wind and hydropower, solid biomass, municipal solid wastes (MSW) and industrial wastes are some of the most important renewable sources for fulfilling these aims. One of the motives to increase the usage of renewable energy sources in Europe is the target set by the European Union (EU), i.e., to obtain 20% of the total energy consumption from renewable sources by 2020 [1]. Targets set by some countries such as in biomass-rich Finland is twice or more than EU's target. In Finland, the

F. Tesfaye (✉) · D. Lindberg · L. Hupa
Johan Gadolin Process Chemistry Centre, Laboratory of Inorganic Chemistry, Åbo Akademi University, Piispankatu 8, 20500 Turku, Finland
e-mail: fiseha.tesfaye@abo.fi

G. Lambotte et al. (eds.), *Materials Processing Fundamentals 2018*, The Minerals, Metals & Materials Series,
https://doi.org/10.1007/978-3-319-72131-6_11

energy produced from solid biomass (wood-based fuels) already accounted for a quarter of the total energy consumption in 2013 [2]. A large share of this energy comes from byproducts of the forest industry including black liquor, bark and sawdust.

Energy sources such as biomass are regarded as CO_2-neutral fuels when handled properly. Therefore, utilizing biomass for energy production is also one of the available alternatives for reducing the use of fossil fuels. Biomass combustion to produce heat and electricity is conducted in large industrial combustors such as fluidized bed boilers, black liquor recovery boilers and grate-fired boilers. Generally, biomass fuels require conversion prior to utilization. Thermochemical conversion methods such as combustion, gasification and pyrolysis are some of the main options. In high-temperature biomass combustion processes, alkali metals in the biomass fuels and the ash fusion behavior are the two major factors contributing to fouling and slagging. These fouling and slagging problems are existing threats for long-term operational availability and costs of biomass power plants [3, 4]. For example, some newly built power plants in Northern China need to shut down and clean their boilers every 20 to 30 days because of the fouling problem in the intermediate-temperature superheaters. Problematic fouling also occurs on or near the bag filters of some boilers [5].

A high concentration of potassium in biomass fuels tend to result in formation of phases and phase mixtures with low melting points. Significant amounts of highly corrosive alkali chlorides in the flue gases emanate from these processes [6]. Slags on the intermediate-temperature superheaters often consist of a considerable percentage of inorganic complex phases such as $K_2Ca_2(SO_4)_3$. The composition of the slag can be KCl + NaCl + $K_2Ca_2(SO_4)_3$ [7]. SiO_2, $CaSO_4$, $CaCO_3$, $K_2Ca_2(SO_4)_3$, KCl, NaCl, $K_2CaMg(SO4)_3$ and $(Na,Ca)Al(Al,Si)_3O_8$ are reported to exist in some boiler ashes [8]. An example of the slag deposits on the heat exchangers is shown in Fig. 1. However, thermodynamic properties of the complex inorganic compounds and their combined effect with the chlorides have been either poorly studied or they are not available. The aim of this work was to contribute thermodynamic data of the alkali sulfates that will help to solve slagging, fouling and high temperature corrosion related problems, which may lead to costly repairs and unplanned shutdowns of the power plants. The current work reviews, updates and contributes new thermodynamic data sets regarding these complex inorganic phases. Reliable thermochemical data regarding the stabilities of the selected compounds are essential for quantitative modeling of the chemical behavior of the complex compounds under different processing conditions. These data sets can also help to develop economically sound and environmentally friendly methods for ash treatments.

Thermochemical Data for Controlling Deposition

A schematic diagram in Fig. 2 shows the role of thermodynamic data in the prediction and control of inorganic materials' deposition on the surfaces of the boiler

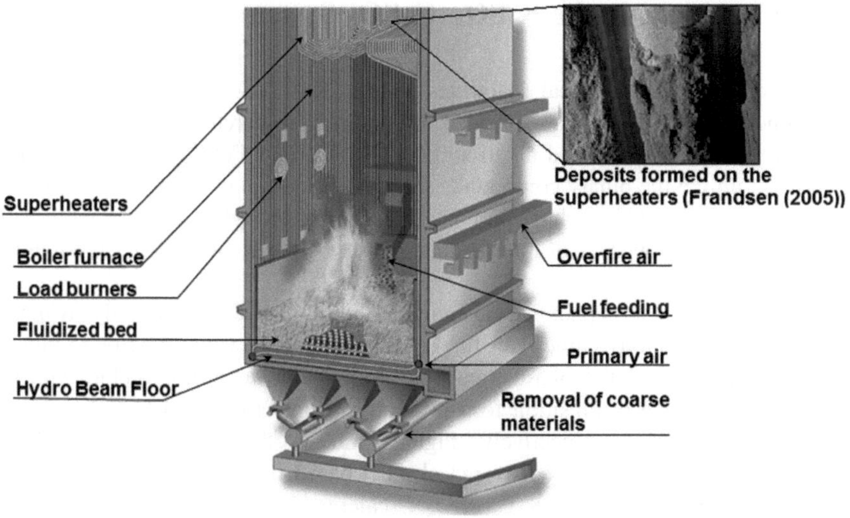

Fig. 1 A bubbling fluidized bed furnace with magnified image showing deposits formed on the surface of super heaters; which causes lower heat transfer, increased corrosion and eventually unscheduled plant shutdown. Image of the deposition was adapted from [9]

Fig. 2 Schematic diagram that shows novel prediction concept for deposition combining advanced fuel analysis, thermodynamics and computational fluid dynamics [10]

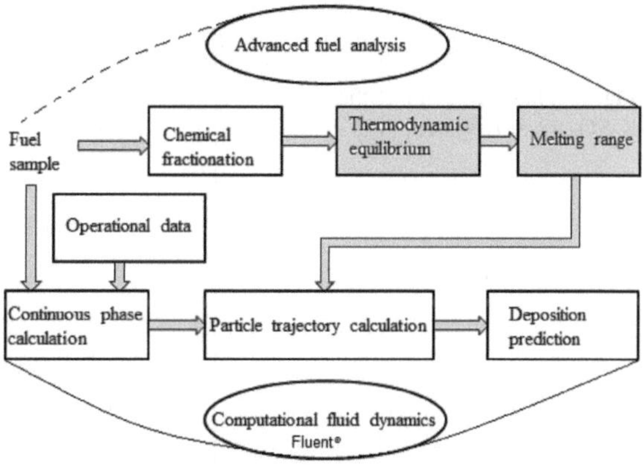

tubes and superheaters. To avoid the intermediate temperature liquid phase formations, the phases and their mixtures that lie along the solidus and liquidus lines must be determined. These data will also help to understand corrosion mechanisms on the superheater surfaces and in the inner surfaces of the flash smelting boiler tubes. In addition to solving problems related to unscheduled plant shut down and additional operations costs, results obtained in this work will also help to improve the materials design of boilers.

Phase Relations in the K_2SO_4–$CaSO_4$ System

Phase relations and thermal stabilities in the K–Ca–S–O systems have been experimentally studied and reviewed by several researchers [11–15]. Crystal structures of the two studied alkali sulfates, K_2SO_4 and $CaSO_4$, are well-defined. However, unlike the binary compounds such as sulfides and oxides thermodynamic properties of the sulfates have not been studied thoroughly. Consequently, the thermodynamic properties of intermediate phases in the K_2SO_4–$CaSO_4$ system are partially unknown.

For instance, heat capacity measurements below 67 °C are the only available experimental thermodynamic data for $K_2Ca_2(SO_4)_3$ [14]. Cao and Boerio-Goates [14] also determined the phase transition (from cubic to orthorhombic) and the enthalpy of the phase transition for the double sulfate to be 201.5 ± 0.1 °C and 2.12 + 0.02 $kJ \cdot mol^{-1}$, respectively. Figure 3 shows the phase diagram of the K_2SO_4–$CaSO_4$ calculated by Rowe et al. [12] based on their experimental observations. According to them, $K_2Ca_2(SO_4)_3$ (calcium langbeinite) is the only intermediate stoichiometric phase in the system. The high-temperature hexagonal K_2SO_4 can dissolve considerable amount of $CaSO_4$, and the low-temperature orthorhombic K_2SO_4 may also dissolve minor amounts of $CaSO_4$. Details on the thermodynamics of the simple sulfates K_2SO_4 and $CaSO_4$, and the double sulfate $K_2Ca_2(SO_4)_3$ are discussed in the subsequent sections.

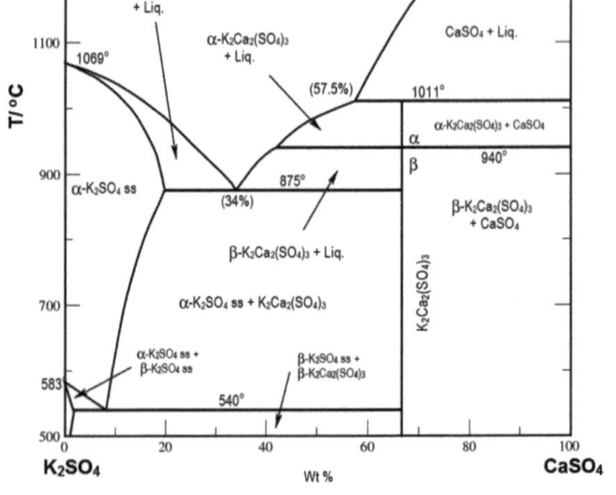

Fig. 3 Phase diagram of the K_2SO_4–$CaSO_4$ system [12]

Experimental Section

Materials Preparation and Thermal Analysis

The provenance and purity of the starting materials used in this study are listed in Table 1. The 99.999% pure K_2SO_4 was analyzed directly. $CaSO_4$–$2H_2O$ was first heated in a protected atmosphere to 400 °C and then cooled to room temperature. Thermal analysis of the hydrophilic $CaSO_4$ was conducted without removing the sample from the protected atmosphere of the thermal analysis equipment. The double sulfate $K_2Ca_2(SO_4)_3$ was synthesized by mixing fine powders of the pure K_2SO_4 and $CaSO_4$ into 1:2 molar ratio. The phase mixture was annealed at 800 °C for about 4 days in a normally locked Nabertherm muffle furnace with good heating control and a continuous flow of N_2(gas). The synthesis took place according to reaction (1):

$$K_2SO_4 + 2CaSO_4 \rightleftarrows K_2Ca_2(SO_4)_3. \quad (1)$$

Thermal analysis of the samples K_2SO_4, $CaSO_4$ and $K_2Ca_2(SO_4)_3$ were done by differential scanning calorimetry (DSC) and thermogravimetry (TGA) method using a NETZSCH STA 449 F1 Jupiter®—Simultaneous TGA-DSC equipment. The calorimeter was calibrated with the melting temperatures and enthalpies of fusion for high purity tin, indium, bismuth, zinc, aluminum and gold. The average measurement accuracies of temperatures and enthalpies of fusion were determined to be ±1 °C and ±1.14%, respectively.

Pt-Rh-crucibles of the same mass were used for the base line (blank run), calibration (sapphire) and sample in all DSC-TGA measurements. Before each DSC measurement the chamber was evacuated and then backfilled with pure Ar(gas) three times. The Ar(gas) was also used as a protective gas. The flow rate of the protective gas was 39 ml/min in all runs. During the runs, initially, the furnace was heated to 40 °C and kept at an isothermal condition for 10 min. Then, the furnace was heated to 1100 °C and cooled to 960 °C, in the case of K_2SO_4, and heated to 1260 °C and cooled to 300 °C, in the case of $CaSO_4$. Weight losses and heat flows during the linear heating and cooling were measured simultaneously. The heating and cooling were at a rate of 10 °C/min for all runs. 19.5 mg of K_2SO_4, 48.74 mg of $CaSO_4$, 60.37 mg of $K_2Ca_2(SO_4)_3$ were used in the independent runs.

Table 1 Provenance and purity of the materials used in this study

Chemical	Mass fraction purity	Source	Form
Ar (argon)	0.99999	AGA (Finland)	Compressed gas
K_2SO_4 (potassium sulfate)	0.99999	Merck (Germany)	Powder
$CaSO_4$–$2H_2O$ (Calcium sulfate dihydrate)	0.99000	Merck (Germany)	Powder

Results and Discussion

The obtained DSC-TGA versus temperature results during heating and cooling of the simple sulfates in the temperature range from 40 to 1260 °C are shown in Figs. 4 and 5. On the heating DSC versus temperature curves, temperatures of phase transition and melting appeared as sharp endothermic peaks in both figures. The integral between the onset and end temperatures of the peaks yielded enthalpies of phase transition and melting. As the TGA versus temperature curves in Figs. 4 and 5 show, weight losses of the samples during heating cycles were negligible. Thus, expected reflection of the peaks on heating DSC curves were also observed on the cooling curves; for example, as solidification temperatures in the case melting temperatures.

Thermochemical Data of K_2SO_4

The first endothermic peak in Fig. 4 with the onset temperature 586.1 ± 1 °C corresponds well with the phase transition temperature of K_2SO_4 reported by Rowe et al. [12], 583 °C. From the integral between the onset and end temperatures of the peak (shaded area in Fig. 4, left side), we determined the enthalpy of phase transition to be $\Delta_{tra}H^{\beta\rightarrow\alpha} = 7.23 \pm 0.1$ kJ·mol^{-1}. The second peak in Fig. 4 with the onset temperature of 1069.2 ± 1 °C is in excellent agreement with the melting temperature of K_2SO_4 reported by Rowe et al. [12], 1069 °C. From the integral between the onset and end temperatures of the peak (shaded area in Fig. 4, right side), we determined the enthalpy of melting to be $\Delta_{melt}H^{\alpha\rightarrow L} = 21.12 \pm 0.3$ kJ·mol^{-1}.

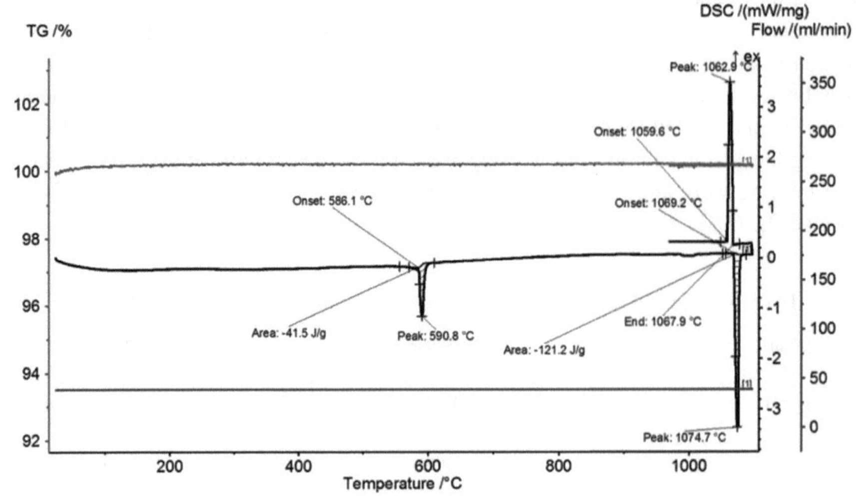

Fig. 4 The DSC-TGA versus temperature curves for K_2SO_4. The experiment was run under the protective Ar(gas) atmosphere using Pt-Rh crucibles

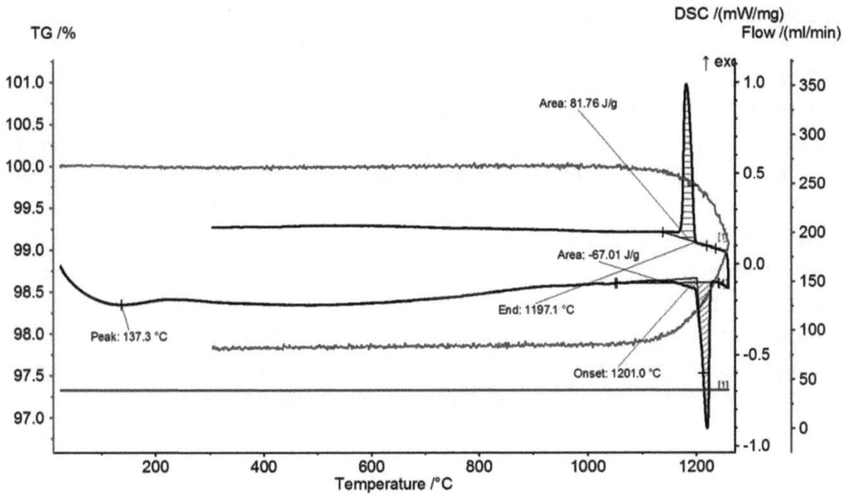

Fig. 5 The DSC-TGA versus temperature curves for $CaSO_4$. The experiment was run under the protective Ar(gas) atmosphere in Pt-Rh crucibles

Thermochemical Data of $CaSO_4$

The starting material $CaSO_4$–$2H_2O$ was annealed at 400 °C before the actual run. As $CaSO_4$ is hydrophilic we kept it in the Ar(gas) atmosphere of the furnace and calculated its actual mass by reducing the weight loss on the measured TGA curve. Then, the actual run was performed and resulted in the DSC-TGA versus temperature curves shown in Fig. 5. The shallow peak with temperature 137.3 °C could be an indication for $CaSO_4$ transition from hydrophilic to hydrophobic. The sharp peak with onset temperature 1201.0 ± 1 °C on the heating curve and end temperature 1197.1 °C on the cooling curve are in agreement with the phase transition of $CaSO_4$ reported by Rowe et al. [12], 1195 °C. Enthalpy of the phase transition was calculated from the peak area between the onset and end temperature to be $\Delta_{tra}H = 9.12 \pm 0.1$ kJ·mol^{-1}.

Thermochemical Data of K_2SO_4–$CaSO_4$

The phase transitions and melting determined for $K_2Ca_2(SO_4)_3$ in this study from the onset temperatures of the three distinct peaks on the DSC versus temperature curve are: 200.1 ± 1 °C, 913.9 ± 1 °C and 1010.6 ± 1 °C. The onset temperature 200.1 ± 1 °C corresponds very well with the phase transition temperature of $K_2Ca_2(SO_4)_3$ determined by Morey et al. [16], 200 ± 2 °C, and Cao and Boerio-Goates [14], 201.5 ± 0.1 °C. The phase transition temperature reported by

Speer and Salje [17], 184 °C, is lower by about 16 °C than the result determined in this study and the result reported by Cao and Boerio-Goates [14, 16].

The second phase transition, $\beta \rightarrow \alpha$, temperature reported by [12], 940 °C, and by [18], 936 °C, are higher than the onset temperature 913.9 ± 1 °C. In the contrary, the incongruent melting temperature they reported $T_{melt}(K_2Ca_2(SO_4)_3)$ = 1011 °C is in excellent agreement with the onset temperature 1010. 6 ± 1 °C measured in this study. The incongruent melting temperature reported by [16], 1009–1013 °C, agree very well with the determined value in this study. The incongruent melting temperature 1004 °C reported by [18] is lower by about 7 °C than the value determined for $K_2Ca_2(SO_4)_3$ in this study. A comparative summary of the phase transition and melting temperatures and the corresponding enthalpies determined from the DSC-TGA versus temperature measurements are compiled in Table 2 together with the available literature values.

Heat Capacity (C_p) Functions

The C_p functions for K_2SO_4 (Eq. 2) and $CaSO_4$ (Eq. 3) were calculated from the measured DSC versus temperature values of the samples and the reference, sapphire. The experimental data were fitted by using the normal format C_p-data fit of HSC Chemistry 6 [19]. The derived functions for both substances agree with the values given in the handbook of Barin [20]. C_p values obtained for $CaSO_4$ in this study and the values reported by Robie et al. [21] are in good agreement up to 327 °C. However, deviations between the two values grow with increasing temperature up to 723 °C.

Table 2 A summary of phase transitions and melting temperatures and enthalpies determined with the DSC versus temperature measurements in the K_2SO_4–$CaSO_4$ system together with the selected literature values

Phase	$T_{tra(1)}$/°C	$T_{tra(2)}$/°C	$\Delta_{tr}H°$/ kJ·mol^{-1}	T_{melt}/°C	$\Delta_{melt}H°$/ kJ·mol^{-1}	References
K_2SO_4	586.1 ± 1	None	7.23 ± 0.1	1069.2 ± 1	21.12 ± 0.3	This work
	583		–	1069	–	[12]
$CaSO_4$	1201.0 ± 1	None	9.12 ± 0.1	–	–	This work
	1195		–	1462	–	[12]
$K_2Ca_2(SO_4)_3$	200.1 ± 1	913.9 ± 1	–	1010.6 ± 1	–	This work
	201.5 ± 0.1	–	2.12 ± 0.02	–	–	[14]
	184	–	–	–	–	[17]
	–	936	–	1004[a]	–	[18]
	–	940	–	1011[a]	–	[12]
	200 ± 2	–	–	1009–1013[a]	–	[16]

[a]Incongruent melting: $K_2Ca_2(SO_4)_3 \rightleftarrows L + CaSO_4$

$$C_p(K_2SO_4)/J \cdot mol^{-1} \cdot K^{-1} = -223.96 + 0.3505255 \cdot T - 6.082803 \times 10^6 \cdot T^{-2} - 7.21025 \times 10^{-5} \cdot T^2 + 5621.6 \cdot T^{-0.5} \quad (2)$$

$$C_p(CaSO_4)/J \cdot mol^{-1} \cdot K^{-1} = 72.45 + 0.112 \cdot T - 4.042 \times 10^5 \cdot T^{-2} + 3.862 \times 10^{-6} \cdot T^2 \quad (3)$$

The C_p functions for $K_2Ca_2(SO_4)_3$ were not calculated in this study. Cao and Boerio-Goates [14] determined C_p for $K_2Ca_2(SO_4)_3$ for up to 67 °C based on results obtained with an adiabatic calorimeter. They reported C_p ($K_2Ca_2(SO_4)_3$) value at 25 °C to be 343.6 ± 0.7 $J \cdot K^{-1} \cdot mol^{-1}$. This C_p-value for the double sulfate agrees with the C_p value calculated by adding C_p values of the simple sulfates (Eqs. 2 and 3) at 25 °C according to reaction (1). Our calculation gave $C_p(K_2Ca_2(SO_4)_3)$ = 334.6 $J \cdot K^{-1} \cdot mol^{-1}$.

Summary and Conclusions

Energy production from renewable sources and the role of thermochemical data to control problems related to fouling and slagging in the combustion processes were discussed. Thermodynamics and phase equilibria of selected phases in the K_2SO_4–$CaSO_4$ system were both reviewed and experimentally studied. Unlike the simple sulfates, thermodynamic properties of the double sulfate $K_2Ca_2(SO_4)_3$ were found to be studied poorly. Thermodynamics of K_2SO_4, $CaSO_4$ and $K_2Ca_2(SO_4)_3$ were studied by applying the simultaneous DSC-TGA analysis technique. The phase transition and melting temperatures determined in this work for the simple sulfates generally agree with the literature values. Furthermore, we determined their enthalpies of phase transition and melting as well as their C_p-values. The melting temperature of $K_2Ca_2(SO_4)_3$ determined in this work, 1010.6 ± 1 °C, also agrees with the literature value of [12, 16], 1011 °C. However, its second phase transition ($\beta \rightarrow \alpha$) temperature determined in this study vary with up to 27 °C from the values reported in [12, 18]. These also support the fact that the phase diagram of K_2SO_4–$CaSO_4$ system need to be reassessed and remodeled.

Acknowledgements The authors are grateful to the Academy of Finland for financial support. This work was made under the project "Thermodynamic investigation of complex inorganic material systems for improved renewable energy and metals production processes" as part of the activities of the Johan Gadolin Process Chemistry Center at Åbo Akademi University.

References

1. EC 2009 A—Renewable Energies Directive, 2009/28/EC of 23 April 2009
2. Official Statistics of Finland (OSF): Energy supply and consumption. ISSN = 1799–7976. 1st Quarter 2013. Helsinki: Statistics Finland [referred: 25 Sept 2015]. Access method: http://www.tilastokeskus.fi/til/ehk/2013/01/ehk_2013_01_2013-06-20_tie_001_en.html

3. Vamvuka D, Zografos D, Alevizos G (2008) Control methods for mitigating biomass ash-related problems, in fluidized beds. Bioresour Technol 99:3534–3544
4. Pettersson A, Amand LE, Steenari BM (2009) Chemical fractionation for the characterization of fly ashes from co-combustion of biofuels using different methods for alkali reduction. Fuel 88:1758–1772
5. Niu Y, Tan H, Wang X, Liu Z, Liu Y, Xu T (2010) Study on deposits on the surface, upstream, and downstream of bag filters in a 12 MW biomass-fired boiler. Energy Fuels 24:2127–2132
6. Karlsson S, Åm L-E, Liske J (2015) Reducing high-temperature corrosion on high-alloyed stainless steel superheaters by co-combustion of municipal sewage sludge in a fluidised bed boiler. Fuel 139:482–493
7. Zhu Y, Niu Y, Tan H, Wang X (2014) Short review on the origin and countermeasure of biomass slagging in grate furnace, 2014, Front. Energy Res 2(7), https://doi.org/10.3389/fenrg.2014.00007
8. Yang Y, Xiao Y, Voncken JHL, Wilson N (2007) Thermal treatment and vitrification of boiler ash from a municipal solid waste incinerator. https://doi.org/10.1016/j.jhazmat.2007.10.116
9. Frandsen FJ (2005) Utilizing biomass and waste for power production—a decade of contributing to the understanding, interpretation and analysis of deposits and corrosion products. Fuel 84:1277–1294
10. Mueller C, Skrifvars B-J, Backman R, Hupa M (2003) Ash deposition prediction in biomass fired fluidised bed boilers—combination of CFD and advanced fuel analysis. Prog Comput Fluid Dyn 3:113–120
11. Rowe JJ, Morey GW, Zen CS (1972) The quinary reciprocal salt systems Na, K, Mg, Ca/Cl, SO_4; a review of the literature with new data, Professional Paper 741, U.S. Govt. Print. Off., 37p. http://pubs.usgs.gov/pp/0741/report.pdf. Accessed Aug 2017
12. Rowe JJ, Morey GW, Silber CC (1967) Ternary system K_2SO_4-$MgSO_4$-$CaSO_4$. J Inorg Nucl Chem 29:925–942
13. Ramsdell LS (1935) An x-ray study of the system K_2SO_4-$MgSO_4$-$CaSO_4$. Am Mineral 20:569–574
14. Cao H, Boerio-Goates J (1994) Heat-capacity measurements and thermodynamic functions of langbeinite-type $K_2Cd_2(SO_4)_3$ and $K_2Ca_2(SO_4)_3$ at temperatures from 10 K to 340 K. J Chem Therm 26:251–259
15. Tian H, Guo Q, Yue X, Liu Y (2010) Investigation into sulfur release in reductive decomposition of calcium sulfate oxygen carrier by hydrogen and carbon monoxide. Fuel Process Technol 91:1640–1649
16. Morey GW, Rowe JJ, Fournier RO (1964) The system $K_2Mg_2(SO_4)_3$ (Langbeinite)—$K_2Ca_2(SO_4)_3$ (Calcium-Langbeinite). J Inorg Nucl Chem 26:53–58
17. Speer D, Salje E (1986) Phase transitions in langbeinites I: crystal chemistry and structures of K-double sulfates of the langbeinite type $M_2^{++}K_2(SO_4)_3$, M^{++}=Mg, Ni, Co, Zn, Ca. Phys Chem Miner 13:17–24
18. Mukimov SM et al (1949) The study of intersection of melt sulfates of Na, K, Mg and Ca, Tr. Inst. Khim. Nauk Uzb. SSR. Inst. Khim. Obsheh. Neorg. Khim., 2:94–112 (in Russia)
19. Roine A et al (2010) HSC Chemistry, version 6, Outotec Finland Oy, Research Center
20. Barin I Thermochemical Data of Pure Substances. Part I & II, VCH Verlagsgesellschaft, Weinheim/VCH Publishers, New York, USA
21. Robie RA, Russell-Robinson S, Hemingway BS (1989) Heat capacities and entropies from 8 to 1000 K of langbeinite ($K_2Mg_2(SO_4)_3$), anhydrite ($CaSO_4$) and of gypsum ($CaSO_4 \cdot 2H_2O$). Thermoch Acta 139:67–81

Waste Lithium-Ion Battery Recycling in JX Nippon Mining & Metals Corporation

Yasufumi Haga, Katsumi Saito and Kazuhiro Hatano

Abstract The JX Nippon Mining & Metals Corporation engages in copper mining, copper smelting and refining, electronic-materials production and used material recycling. The characteristic of its recycling business is applying "Zero-Emission Policy" without landfill, with operating a unique metal recovering system consists of pyro- and hydro-metallurgical processes, and with employing efficient scrap treatment system utilized from its copper smelting technology. In 2010, we constructed a pilot plant for recovering valuable metals from waste lithium-ion batteries and its cathode materials. The process consists of leaching, solvent extraction and electro-winning to recover electrolytic cobalt and electrolytic nickel as main products, and lithium and manganese carbonate as byproducts. We have changed its mind to processing mainly the waste batteries because of the difficulty of collecting waste cathode materials. This paper describes the technical changes and development, and also shows the recent operating conditions of the plant.

Keywords Lithium-ion battery · Battery recycling · Zero emission

Introduction

JX Nippon Mining and Metals Corp. (JX-NMM) engages in recovering valuable metals and detoxifying industrial wastes. The company performs business with superior technologies of combustion, extraction and refining on non-ferrous metals. JX-NMM started nickel and cobalt smelting and refining business with solvent extraction process 40 years ago.

Lithium-ion batteries (LIB) has become indispensable in our lives. LIB contains some valuable metals such as manganese, cobalt, nickel and lithium. The demand for LIB has spread widely and its production volume has increased rapidly.

Y. Haga (✉) · K. Saito · K. Hatano
JX Nippon Mining & Metals Corp., 1 Wakamizu-cho, Tsuruga city, Fukui 914-0027, Japan
e-mail: haga.yasufumi@jxgr.com

G. Lambotte et al. (eds.), *Materials Processing Fundamentals 2018*,
The Minerals, Metals & Materials Series,
https://doi.org/10.1007/978-3-319-72131-6_12

We decided to commence LIB recycling business considering the increase of waste LIB.

We completed the construction of a plant to recover valuable metals from LIB in 2010 with the financial support of the Ministry of Economy, Trade and Industry (METI). One of our main raw materials was waste cathode materials which contain some impurities, and we have changed the raw materials, and have made the process to be acceptable for a larger amount of impurities.

This paper describes technological developments to control impurities level in the process and recent operational conditions of the plant.

Background to LIB Recycling

Nikko Nickel Cobalt Refining Co., Ltd. was a subsidiary company of JX-NMM and started its nickel and cobalt business in 1975. The company was the first company which actually applied PC-88A for cobalt separation and developed an effective nickel and cobalt extracting process. They developed and piled many knowledge of hydrometallurgical refining process, especially on the field of solvent extraction. JX Nippon Tsuruga Recycle Co., Ltd. started the thermal treatment business for treating waste LIB from mobile phones in 2000. We have had enough experience on some technologies over burning, crushing and separation and we applied those technologies for the waste LIB processing. Then, we started test running to make sure for our process to be feasible at the Tsuruga Plant in 2010.

LIB Recycling Process

Figure 1 is a basic flowsheet for the Tsuruga plant. This process consists of leaching step, solvent extraction step and each metal recovery step. The feed materials are leached in acidic solutions. Manganese, cobalt, nickel and lithium are separated from the pregnant solution of its leaching step sequentially and elements are condensed into mother solution of its recovery step through each solvent extraction step. Cobalt and nickel are recovered by electrowinning and manganese and lithium are recovered as carbonates.

The Development of the Process

At first, we expected that our main raw materials were ingredients of cathode materials of LIB from a manufacturing plant. The waste cathode are composed of mainly manganese, nickel, cobalt and lithium. It also contains a little amount of copper and aluminum. They should be removed before solvent extraction steps

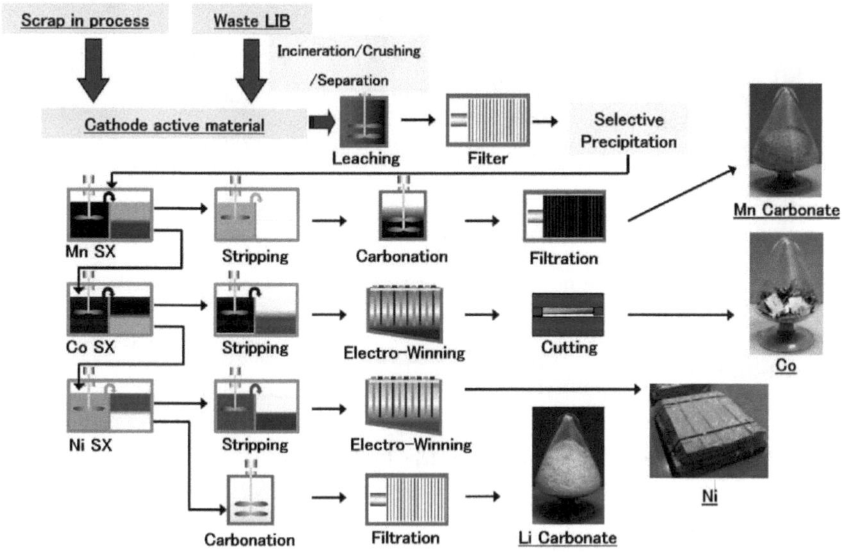

Fig. 1 Flowsheet of LIB recycling process at Tsuruga plant

because they have negative effects on solvent extraction operation and on the quality of products. We employed a mixed organic solvent to remove aluminum and copper together with manganese at the manganese extraction step after some examinations.

When we started processing the waste cathode, it was found that to collect waste cathode materials became difficult because the amount of waste cathode materials from a manufacturing plant were limited. The competition to collect pure materials became keen so that we focus on LIB scrap. There were some problems about impurities to be solved, but the amount of LIB scrap on the market were larger than the waste cathode materials. We decided to change the major feed from waste cathode materials to LIB scraps.

We should remove many other things, like casing, wire, connector and so on, from LIB scrap to recover cathode materials. We incinerate at a stationary furnace to evaporate and burn out organic electrolyte followed by crushing and screening its clinker. Fluorine in the organic electrolyte is precipitated and removed in JX Nippon Tsuruga Recycle Co., Ltd. After crushing and screening, the clinker is separated into coarse material and fine material. The coarse material is sent to our smelting section in Kyushu to recover the copper, and the fine materials which is mainly cathode material is sent to the Tsuruga plant. Figure 2 shows the flowchart of crushing and screening steps for LIB scrap.

Table 1 shows the chemical compositions in raw materials. The crushed cathode materials from LIB scrap has too much impurity to process the same way we used before. We developed a new process to adjust impurity and confirmed that the most of impurities were removed simply.

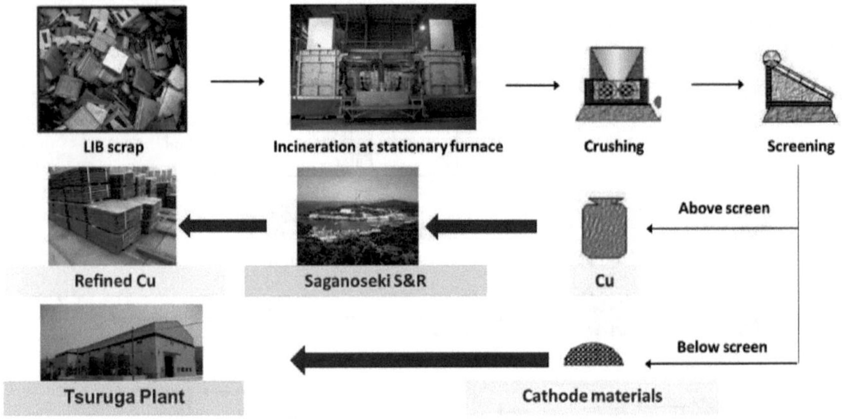

Fig. 2 Flowsheet of crushing and screening steps for LIB scrap

Table 1 The differences of composition in raw materials

Raw materials	Typical composition (%)						
	Mn	Co	Ni	Li	Cu	Al	Fe
1. Waste cathode materials	20	20	20	7	0.01	0.1	0
2. Crushed cathode materials from LIB scrap	0.3	35	0.5	4.0	2.0	5.5	0.1

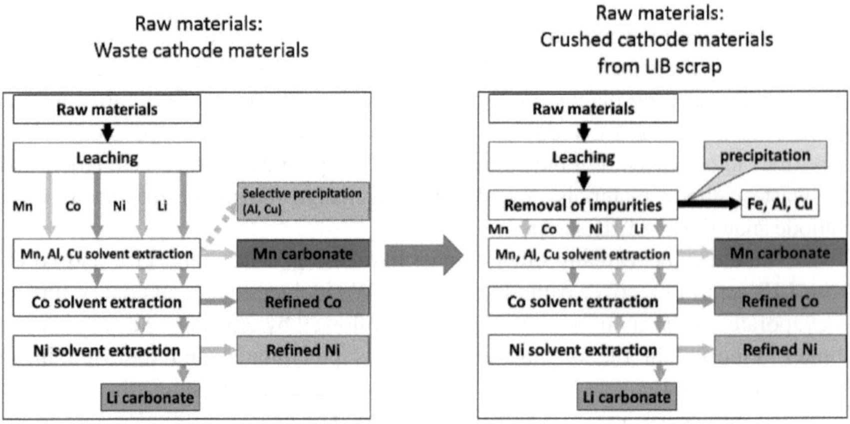

Fig. 3 Flowsheets for each raw materials

Figure 3 is flowsheets for each raw materials. We put removal step before solvent extraction step. First, we control redox potential in acidic solution to prevent leaching out of copper. Then, we adjust pH to precipitate aluminum as hydroxide. After filtration, iron is removed by adjusting pH and redox potential.

Table 2 Leaching rate and concentrations after removal of impurities

Raw materials: crushed cathode materials from LIB scrap	Co	Li	Cu	Al	Fe
Leaching rate (%)	>90	>90	<0.1	<10	<0.1
Concentrations (g/L)	25–30	3.5–4.0	<0.01	<0.4	<0.001

Table 2 shows the reaching rate and concentrations after removal step. The concentration of aluminum after removal step is 400 mg/L. The leaching rate of cobalt lower if we try to precipitate all aluminum at removal step, therefore we leave some amount of aluminum in leaching solution and the residual aluminum is removed at solvent extraction step. We have operated this process steadily and efficiently.

Conclusion

The JX-NMM operates the test plant at Tsuruga plant to recover valuable metals from waste LIBs. We developed our process to be able to accept many impurities as impurity levels were increased according to the change of raw materials. And now, we can process feed materials containing much impurity.

The principle purpose of our plant is zero emission principle. The coarse material and the residue of lower material of waste LIB are sent to our smelting section to recover copper.

Recovery of Platinum Group Metals Out of Automotive Catalytic Converters Scrap: A Review on Australian Trends and Challenges

Maryam Ghodrat, Pezhman Sharafi and Bijan Samali

Abstract The Platinum Group Metals (PGMs) are of substantial technological prominence. They are also extremely rare, because of their low natural existence and their complicated extraction and refining process. To meet the future demand and preserve resources, it is essential to process end-of-life platinum-containing materials, such as catalytic converters. PGMs recovery from catalytic converters scrap commonly carried out by pyro/hydrometallurgical processes that involved thermal treatment followed by leaching and solvent extraction. This paper reviews current methods in used in the recovery of PGMs out of waste catalytic converters in Australia and discusses some of the key factors and opportunities in improving the existing methodologies. Emerging trends that are likely to affect the current or future PGM recovery are also explored.

Keywords Catalytic converter · Platinum group metals · Waste recovery Australia

Introduction

The Platinum Group Metals (PGMs) comprise platinum (Pt), palladium (Pd), iridium (Ir), osmium (Os), rhodium (Rh) and ruthenium (Ru) [1]. The elements of most commercial significance are platinum, palladium and, to a lesser degree, rhodium. The PGMs properties of commercial importance are their resistance to corrosion and oxidation, high-melting points, electrical conductivity and catalytic

M. Ghodrat (✉) · P. Sharafi · B. Samali
School of Computing, Engineering and Mathematics, Centre for Infrastructure Engineering, Western Sydney University, Sydney, Australia
e-mail: m.ghodrat@westernsydney.edu.au

P. Sharafi
e-mail: p.sharaf@westernsydney.edu.au

B. Samali
e-mail: b.Samali@westernsydney.edu.au

G. Lambotte et al. (eds.), *Materials Processing Fundamentals 2018*,
The Minerals, Metals & Materials Series,
https://doi.org/10.1007/978-3-319-72131-6_13

activity in the chemical, electrical, electronic, glass and motor vehicle industries. The automotive industry is the principal customer of PGMs. Autocatalytic converters in vehicles require palladium, rhodium, and platinum to convert exhaust emissions to water and carbon dioxide [2, 3]. Concentrations of precious metals in many End-Of-Life (EOL) products are much higher than those found in ore deposits [4]. Furthermore, the product represents a different and usually a much simpler matrix for metal separation, with less waste being generated, and a reduction in energy requirements to recover the target metal(s) as there is no need to remove large quantities of gangue material [5]. An automotive catalytic converter contains approximately 2000 g/mt of PGMs in the ceramic block, compared to average PGMs concentrations of <10 g/mt in most PGM mines [6]. Considering the high environmental impact of primary production of precious metals arising from low ore concentrations, difficult mining conditions, high energy and water use, high chemical consumption, large waste generation, and other factors, recovery of metals from EOL products is appealing. The key to high overall recycling rates is collection. Automotive catalytic converters are collected at a relatively high rate due to their ease of retrieval and intrinsic value. The global recycling rates for Pt, Pd, and Rh from automotive catalytic converters range from 50 to 55% [7, 8], indicating that around half of the catalytic converters produced are collected for recycling. The main focus of this paper is to review existing trends, and recycling strategies for platinum group metals in Australia.

PGMs State in Australia

PGMs are categorised as critical commodities in Australia. The critical commodities are defined as metals, non-metals and minerals that perform an essential economic function but are subject to a high risk of supply. The supply risk factors could be related to geological scarcity, a lack of substitution or recycling, geopolitical instability, a concentration of production and processing in particular countries or companies, a lack of large-scale markets, production only as a by-product and limited methods of recovery. Elements assessed as having category one (high critically) resource potential in Australia are rhodium, manganese, indium, platinum, rare earth elements and palladium. Table 1 compares demand in 2006 with projected demand in 2030, for platinum, palladium and copper in the emerging technologies sector (European Commission, 2010). In 2006 the demand from emerging technologies generally comprised only a small to moderate fraction of global production. Notably, projected annual demand in 2030 for indium, germanium and platinum all exceed current annual production (European Commission, 2010). Projected demand in 2030 is proportionately lower for silver, cobalt, palladium, titanium and copper. The data presented in Table 1 demonstrates that all of the elements listed could experience large to very large growth in demand by 2030.

Critical commodities assessed as having category one (high) resource potential in Australia are chromium, cobalt, copper, nickel, platinum-group elements (PGE),

Table 1 Comparison between production and demand for selected commodities, 2006 and 2030 (European Commission, 2010) [9]

Raw materials	Production 2006 (tonnes)	Demand for emerging technologies 2006 (tonnes)	Demand for emerging technologies 2030 (tonnes)	Indicator 2006	Indicator 2030
Indium	581	234	1911	0.4	3.29
Germanium	100	28	220	0.28	2.20
Platinum	255	Very small	345	0	1.35
Silver	19,051	5342	15,823	0.28	0.83
Cobalt	62,279	12,820	26,860	0.21	0.43
Palladium	267	23	77	0.09	0.29
Titanium	72,211,000 (ore concentrate)	15,397	58,148	0.08	0.29
Copper	15,093,000	1,410,000	3,696,070	0.09	0.24

rare-earth elements (REE), and zirconium and critical commodities assessed as having category two resource potential in Australia are antimony, beryllium, bismuth, graphite, helium, indium, lithium, manganese, molybdenum, niobium, tantalum, thorium, tin, titanium, and tungsten. Supply of PGMs is dominated by South Africa and Russia; Australia has category one potential for discoveries of new resources due to favourable geology. Demand is expected to grow due to uses in vehicles, clean energy technologies and desalination. Table 2 presented the assessment of Australian potential to supply some of the critical commodities including PGMs.

Platinum-Group Metals Supply and Demand

According to figures published by Matthey [10], gross demand for platinum fell marginally in 2012 to 8.05 million ounces (Moz). Demand for platinum in auto-catalyst applications rose slightly in 2012 to 3.24 Moz. The demand for platinum in jewellery increased by 305,000 oz (oz) to 1.57 Moz while the gross industrial demand for platinum fell by 405,000 oz in 2012. Net identifiable physical demand for platinum in the investment sector reached to 455,000 oz in 2012, about 5000 oz lower than in 2011. Gross demand for palladium increased by 16% to 9.90 Moz in 2012. Gross demand for palladium from the auto-catalyst sector in 2012 reached an all-time high level of 6.62 Moz as a result of greater use of palladium in light duty gasoline vehicles in China and another rise in the ratio of palladium to platinum in auto-clasts for European diesel vehicles [11]. Demand for palladium in industrial applications fell by 100,000 oz to 2.365 Moz, while demand for jewellery fell by 60,000–445,000 oz but net physical investment demand switched from a negative of 565,000 to a positive of 470,000 or an increase of more than 1 Moz to 470,000 oz in 2012 [12]. A snapshot of the aforementioned data is also given in Fig. 3.

Table 2 Comparison of Australian potential to supply PGMs and other critical commodities

Commodity	Criticality (Score)	Australian production (2011) as % of global	Economic resources World; Australia
Rare earth elements	Category one (29)	0%	114,000 kt; 2070 kt
Platinum group elements (Including Pt and Pd)	Category one (22)	0.1%	66,000 t; 4.7 t
Cobalt	Category one (21)	4%	7334 kt; 1204 kt
Nickel	Category one (13)	12%	76,000 kt; 20,400 kt
Copper	Category three (2)	6%	690,000 kt; 86,700 kt
Indium	Category one (26)	n.a.	n.a.
Titanium	Category two (10)	15% ilmenite, 62% rutile	Ilmenite: 1235 Mt; 189 Mt Rutile: 51.3 Mt; 27.2 Mt
Germanium	Category two (8)	0%	n.a.

PGM Resources

Australia's resources of PGMs remained unchanged at 4.7 tonnes in 2012. Western Australia (WA) and the Northern Territory (NT) hold all of Australia's PGM Resources but additional resource in other States may be established [13]. However, the resource of PGMs in individual deposits within different State of Australia is often unrecorded, resulting in the overall unknown distribution of PGM resources [14]. Total identified resources of PGMs are about 276 tonnes. Of this amount, deposits that have only PGM resources account for about 51% of the total resources, although all of Australia's production is as a by-product from PGM resources associated with nickel sulphide deposits in Western Australia [15]. Currently, 400 kg of the published PGM resource is accessible for mining while the balance of 4.3 tonnes occurs within national parks. The reason for this low Accessible figure for PGMs in Australia is that PGM resources are generally not reported by nickel-cobalt producers where PGMs are a by-product of nickel mining [16].

PGM Production

Australia's PGM production (platinum and palladium) in 2012 amounted to 706 kg, which was very minor when compared to world standards [17]. The production was exclusively from nickel sulphide deposits hosted by Archean komatiitic rocks in the

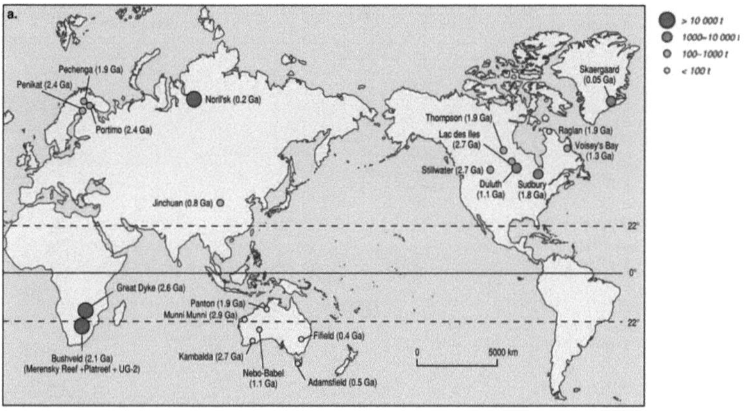

Fig. 1 Platinum-group-metals deposit and platinum-group-metals-bearing nickel deposit [10]

Yilgarn Craton of Western Australia [9]. Australia's resource potential is high for discovery of major platinum-group element and chromium deposits [9]. Areas of known and inferred resource potential for these elements are shown in Fig. 1.

Market Survey for PGM Recycling

The use of vehicles equipped with a catalytic converter resulted in the increase of world PGM demands. The market conditions of PGM, which influence their recycling, are rather volatile [18–20]. A market review was carried out and its outcome showed that, PGMs are mined almost exclusively in particular areas of Australia [21]. As their market operates under the influence of variable conditions, PGM prices show large variations; and some PGMs such as rhodium are not negotiable in stock market [22]. Rhodium production is associated with a concurrent production of substantial platinum amounts, which is undesirable [23]. Almost 81% of the rhodium mined is used for the manufacture of catalysts, compared with only 42% of platinum and only 8% of palladium [24]. The lack of considerable rhodium resources and the inability to satisfy the increased demands, which arose from the use of catalytic vehicles, resulted in the 10-fold increase of its price during the last decade [25–27].

Influence of Automotive Catalysts on PGM Demand and Prices

As shown in Fig. 2b, d, f, the global demand for platinum (Pt), palladium (Pd) and rhodium in auto-catalyst production between 2004 and 2013 had the highest levels among the other applications. The largest increase in demand for auto-catalyst

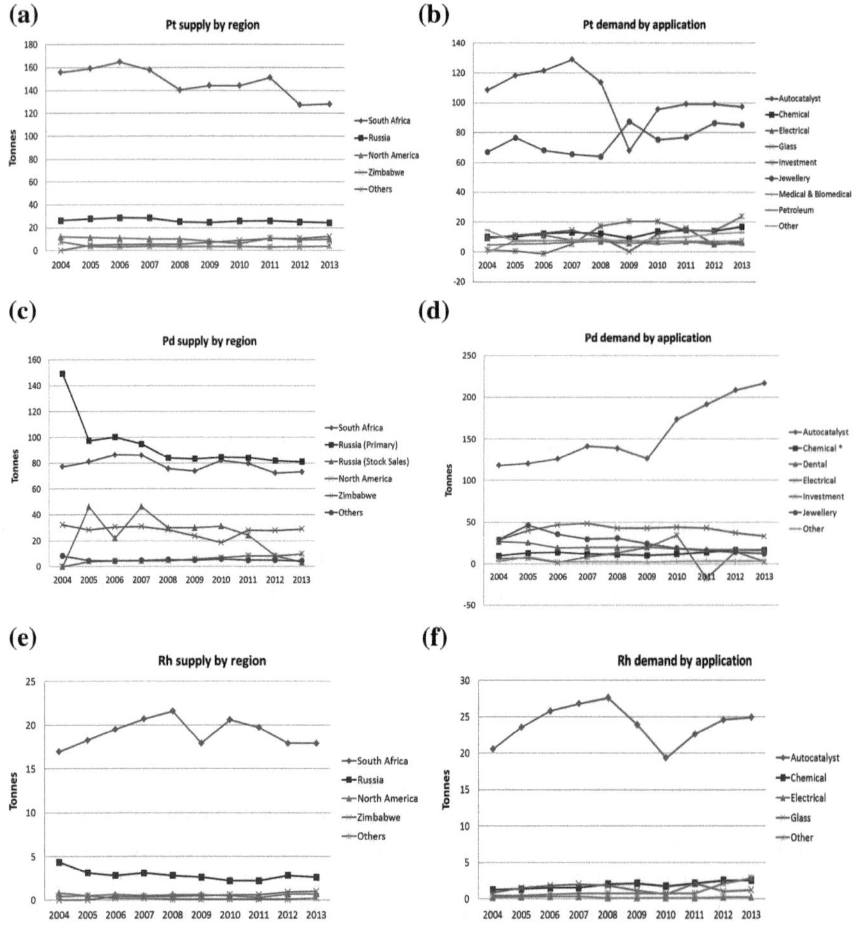

Fig. 2 Platinum, palladium, and rhodium supply by region (**a**, **c**, **e**) and demand by application (**b**, **d**, **f**) since 2013 [10]

(from 2009 onward) has come from Pd, which also had a considerable effect on the price of that metal (Fig. 2b). A sharp decline in the annual demand for Pt and Rh started roughly in 2008 and continued into 2010, and was mainly caused by the effect of global financial crises and falling stocks for these metals. The declining trend for Pd is slower as Pd withdrew from internal stocks within the automotive industry in 1999/2000. Since the use of automotive catalysts slowly started in the second half of the eighties, the PGM-demand curve shows a steady increase. Due to the recent strong market share of Diesel cars, the Pt demand for catalysts in Australia has been raised considerably. As can be seen from Fig. 2a, e, f, a strong discrepancy exist between PGMs supplies with South Africa being the largest supplier for Pt and Rh and Russia for Pd.

PGM Recycling Potential

With an average car life in Australia of over 15 years [28], the largest portion of the recyclable PGMs used for auto-catalysts is still on the roads. Total accumulative recycling quantities from end-of-life catalytic converters up to 2002 are estimated as 260 t globally [29]. In this context, the current car parks can be referred to as the "mine above ground". Making use of this mine through recycling of car catalysts offers a valuable source for future PGM supply not only from an economical point of view but also in terms of securing sustainable supply of PGMs and decreases today's heavy dependence on primary resources. Nevertheless, the share of secondary supply of PGMs for auto-catalysts is still insignificant [30]. The main reason might be that what we can recycle today from scrap catalysts commonly comes from catalyst production more than 10–12 years ago and at that time total PGM usage for this application was much lower [31]. So in a dynamically surging market, the recycling share seems to be insignificant. Extensive PGM-losses occur due to system leaks (e.g. scrap-car exports out of Australia) and inadequate operating of the recycling chain. However, with growing market maturity the recycling share has been improved lately. Figure 3 compared the global recovery of palladium (Pd) and platinum (Pt) from spent autocatalysts during 2008–2012. As it can be seen from Fig. 4, the global recovery of palladium increased by 4.5%, from 2008 to 2012, however for platinum the trend is quite variable in that the Pt recovery decreased sharply during 2008–2009 then slightly increased to the level of 2011.

PGMs Recovery Techniques

Selection of an efficient and economic recycling technique for PGM recovery out of waste catalytic converter mainly depends on catalytic supporter and PGM loading content and other base metals involved in the recovery process [30]. The spent

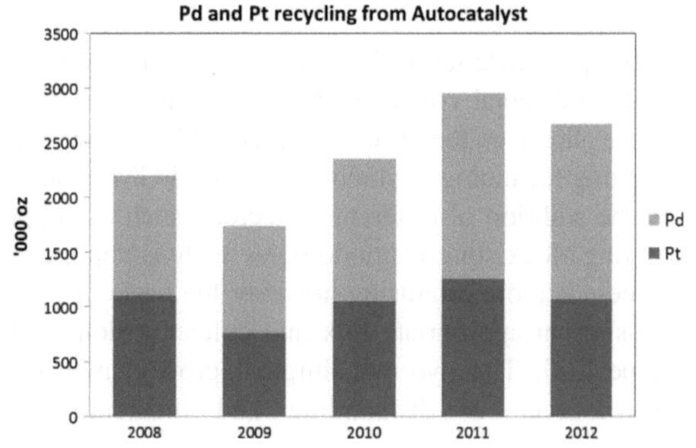

Fig. 3 Recovery of palladium (Pd) and platinum (Pt) from spent autocatalysts [32]

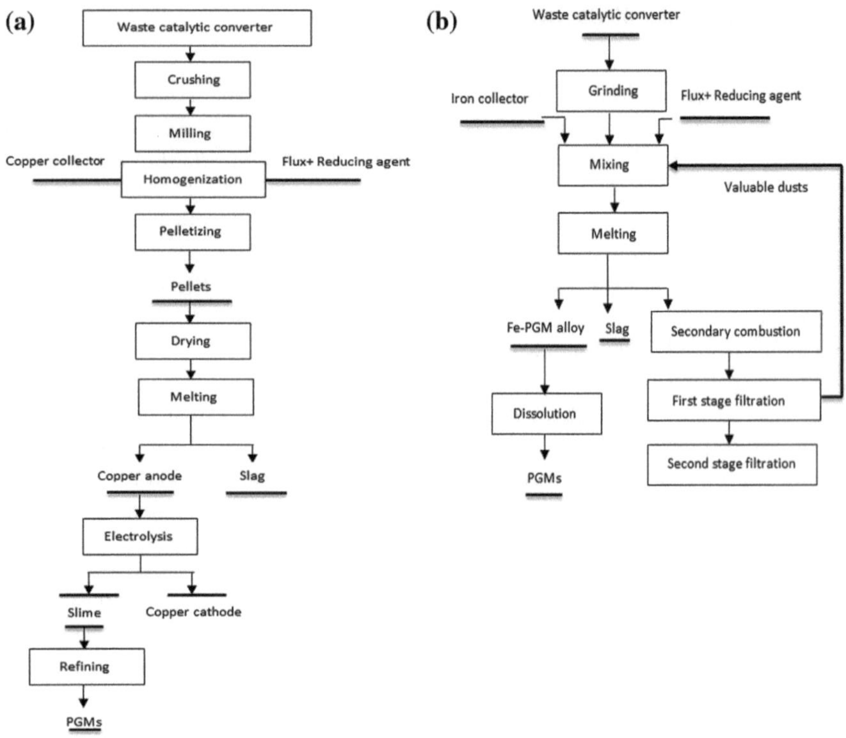

Fig. 4 Flow chart of recovery of PGMs from waste catalytic converter via copper collection (**a**), Iron collection (**b**) [33]

catalytic converters are first separated, shredded, ground and then processed by either pyro/hydrometallurgical or a combination of both.

Pyrometallurgical Route

In the pyrometallurgical route, waste catalysts are melted by adding flux components and metal collectors at high temperatures to form PGM-containing alloys. These alloys are then used to extract PGMs through proper refining techniques. The smelting technology, which produces a PGM-containing alloy, is largely controlled by the addition of different collectors, such as copper, iron, lead, matte, etc. [34]. Among all existing methodologies such as vaporization and sintering, the smelting process is more promising however the basic operation can be further upgraded by utilising an appropriate flux and collector along with suitable pre-treatment technique [33]. The pyrometallurgical process may consist of primary pretreatment phases such as dismantling and incineration of nonmetallic components, calcination, or reduction. After that, high-temperature incineration is use in which an appropriate collector (base metal) such as lead, copper, nickel, etc., collects the PGMs and produces PGM-enriched alloy. Simultaneously the catalyst carrier, such as alumina, is melted with suitable fluxes to acquire a low-viscosity liquid slag [35]. While PGM separates from the slag the PGM containing alloy will be sent for purification. Selecting the right collector is critical in PGM recovery through the pyrometallurgical route. The important parameters in the collector selection include melting points, chemical properties of the collector and common solubility. Usually copper, lead, iron, matte and waste printed circuit board which contained the aforementioned metals are suitable as a collector for smelting of catalytic converter scraps. Other metals such as calcium, magnesium, and zinc generally lead to poor total recoveries of PGMs (<80%). Figure 4a shows copper collection process which consists of crushing, milling, homogenization, pelletizing, drying, melting, electrolysis, and refining [35]. The ceramic carrier of PGMs in waste catalyst converter generate a slag with less than 0.2 wt% copper after enrichment of PGMs into the molten copper. The created slag is then poured out of the furnace for recycling of valuable components [33]. The molten copper contained PGM is casted in anode form about 1 wt% PGMs for electrolysis. A slime with around 20–25 wt% PGMs is collected on the anode after electrorefining. High-purity copper (99.99%) is caught on the cathode. Figure 4b shows flow chart of recovery of PGMs from waste auto catalytic convertor through iron collection. Iron is a low-cost collector with high chemical affinity for PGMs to create solid solution [36]. In this method, crushed waste auto catalysts blended with iron ore. a reducing agent such as coke, and flux such as CO, are melted together in an electric arc furnace at a temperature of 1450 ° C [34]. The alloy and slag can simply separate because of the large difference between densities of the obtained Fe-PGMs alloy and slag [37, 38].

Hydrometallurgical Route

In the hydrometallurgical route, PGMs are recycled by dissolving in aqua regia and acids, including hydrochloric acid, nitric acid and sulfuric acid, along with oxygen, iodine, bromine, chlorine or hydrogen peroxide [30]. The main process of recovering PGMs from spent catalyst by the hydrometallurgical process is shown in Fig. 5. In general, the waste auto catalytic converter is pre-treated under some settings to eliminate organic substances and carbon deposition on the surface, which is then followed by the hydrometallurgical treatment [30]. Finely-ground waste catalytic converter is dissolved by sulfuric acid in this manner and Pt and Pd are concentrated in the leached residue. According to Dong [30], a few dissolved PGMs can be recovered by a cementation process using aluminum powder as reducing agent. Some researchers have also investigated the recovery of PGM out of spent auto exhaust catalytic converter through pressure leach by sulfuric acid. In Australia, the prevailing hydrometallurgical route in use is through leaching of Pt and other PGMs. Leaching is usually carried out using different liquid mediums

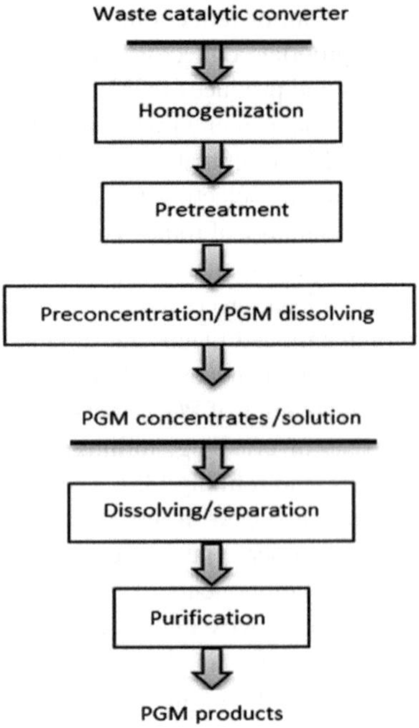

Fig. 5 The principal flow sheet of PGM recovery by hydrometallurgical process [30]

such as sulfuric acid, hydrochloric acid, nitric acid, sodium cyanide, chloride or iodide solutions to selectively extract PGMs from spent catalytic converter. The oxidizing agents are then added to improve the leaching efficiency of PGMs.

The Method in Use

The dominant methodology in use in Australia for recovery of PGMs from secondary sources consists of a combination of pyrometallurgical, hydrometallurgical and electrolytic steps within one process. The PGMs contained wastes are mixed with a collector metal, such as Cu and smelted in a shaft furnace to extract the PGMs from the waste material by forming an alloy with the added metal (Fig. 6). The resulting Cu alloy contains up to 20 wt% PGMs [34]. Slag-forming agents are fed into the furnace to remove oxidic impurities. Afterwards, the Cu-PGM alloy is cast into anodes and subjected to electrolysis, through which PGMs are accumulated in the sludge. Alternatively, the alloy can be leached with sulfuric acid, with the PGMs remaining completely in the solid, undissolved residue. Subsequently the residue is treated comparably to original raw materials, starting with aqua regia leaching [34].

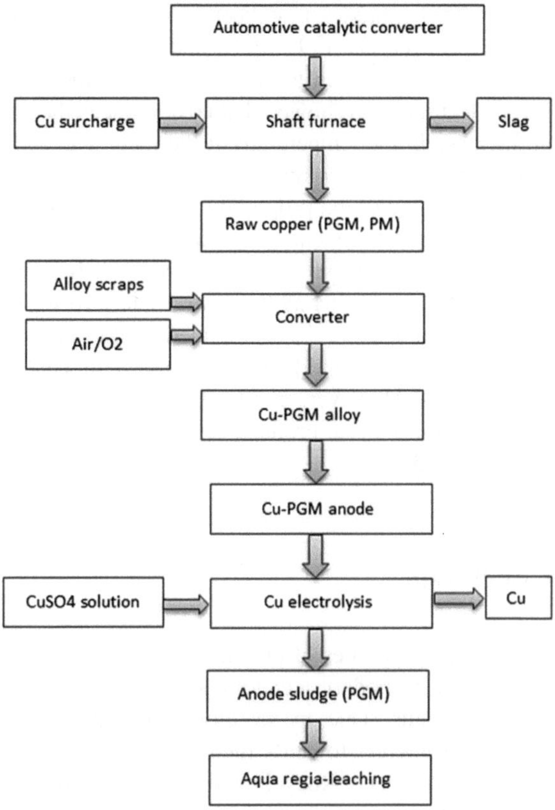

Fig. 6 PGM recovery from end-of-life catalytic converter originating from copper production

Proposal for an Improved Process

A potential method for recovery of PGMs contained in waste automotive catalytic converters is proposed in this section. The method entails pyrolytic concentration in an electric arc furnace followed by a subsequent electrochemical separation. We believe the approach holds promising recovery rate and is relatively simple to implement. The process starts with the melting of crushed automotive catalysts in an electric arc furnace to obtain a collector-metal with PGM content of up to about 30%. The collector-metal is then subjected to electrochemical separation. The charge to the electric arc furnace consisted of approximately 75% crushed catalytic converter, 16% lime, 5% finely dispersed coke and 4% collector-metal (iron-ore concentrate). The holding temperature inside the furnace is 1600 °C. Collector-metal in the form of cast-iron chips placed on the bottom of the furnace then crushed catalysts mixed with finely burnt lime poured into the furnace. The collector-metal added to the composition of the coke. This led to the PGMs combining with the collector-metal before the coke melted. The PGMs became coarser in the process and quickly settled onto the bottom of the furnace after the charge melted. This approach to collection shortens the process, while the introduction of iron-ore concentrate (99% FeO) and a reducing agent into the coke makes it possible for the collection to take place at the molecular level rather than the droplet level. Heat analysis showed that coke reduces the amount of recovered metal due to an increase in the temperature of the melt. The addition of a collector-metal to the coke decreases the volume of dust-gas emissions and increases PGM yield. The main advantage of the proposed method is that the formation of the collector-metal shortens the subsequent processing time and the chemical composition of the collector-metal can be reliably determined. Introduction of nitric acid is the next stage of the process which leads to a selective separation of PGMs. The amount of acid required is about 5% of the feed inlet to the furnace. This low consumption of acid makes the method more efficient and cost-effective than other acid-based technologies. However, similar to all other chemical conversion processes, this stage generates waste products and requires the use of some other waste treatment equipment [39].

Conclusion

The unique properties of platinum group metals and a lack of alternatives for their application in modern technologies, especially automotive industry underpin their strategic status. Due to scarcity and high value of these metals, there is an increasing interest towards their recovery from wastes, such as end-of-life auto catalytic converter. Recycling of metals from end-of-life products can augment supply in the future, given a sufficient price incentive and/or regulation. This paper presented an overview of PGMs resource and production in Australia. The supply and demand for PGMs and their potential recycling rate has also been discussed. The main processes for the recovery of PGMs from end-of-life automobile catalytic converter namely hydrometallurgical and pyrometallurgical have been explored and a new methodology for PGMs recovery has been proposed by the inclusion of a reducing agent and an ore-bearing material to the coke. The use of iron-ore concentrate and finely dispersed coke makes it possible to collect PGMs throughout the volume of the coke and maximize PGMs recovery. While recycling is not sufficient to alleviate all challenges associated with considerable growth demands, it can play a valuable role in improving the security of metal supplies. Finally, a better understanding of the risks associated with PGMs recycling is a critical factor for efficient recovery. The inclusion of the collector-metal such as iron-ore concentrate in coke makes it possible to shorten the holding time for the melt and reduce the volume of dust-gas emissions which is the foremost environmental benefit for the proposed method. It is equally important for the minimisation of environmental impacts arising from new recycling facilities, and for defining a comprehensive base to compare PGMs recycling with primary production. The values of the existing PGMs recovery facilities, based on the economic, technical and environmental factors, suggest that recycling projects are the most advantageous.

References

1. Suoranta T, Zugazua O, Niemelä M, Perämäki P (2015) Hydrometallurgy 154:56
2. Barakat MA, Mahmoud MHH (2004) Hydrometallurgy 72:179
3. Hoffmann JE (1988) JOM 40:40
4. Zhang J, Everson MP, Wallington TJ, Field FR III, Roth R, Kirchain RE (2016) Environ Sci Technol 50:7687
5. Froehlich P, Lorenz T, Martin G, Brett B, Bertau M (2017) Angew Chem Int Ed Engl 56:2544
6. Angelidis TN, Sklavounos SA (1995) Appl Catal A Gen 133:121
7. Moskalyk RR, Alfantazi AM (2003) Miner. Eng. 16:893
8. Shelef M, McCabe RW (2000) Catal Today 62:35
9. Skirrow RG, Huston DL, Mernagh TP, Thorne JP, Duffer H, Senior A (2013) Critical commodities for a high-tech world: Australia's potential to supply global demand. Geoscience Australia Canberra
10. Mathey J (2013) Platinum 2013. http://www.platinum.matthey.com/media/1614079/platinum_2013.pdf
11. Chegwidden J, Kingsnorth DJ (2011) Present Inst Anal Glob Secur
12. Panayotova M, Panayotov V (2012) Univ Min Geol
13. Australia's Identified Mineral Resources (2016). http://www.ga.gov.au/scientific-topics/minerals/mineral-resources/platinum#_edn1. Assessed 15 July 2017
14. Jaireth S, Hoatson DM, Miezitis Y (2014) Ore Geol Rev 62:72
15. Laznicka P (2014) Ore Geol Rev 62:259
16. Golev A, Scott M, Erskine PD, Ali SH, Ballantyne GR (2014) Resour Policy 41:52
17. Mudd GM (n.d.) Platin Met Rev 56:2
18. Massari S, Ruberti M (2013) Resour Policy 38:36
19. Schlinkert D, van den Boogaart KG (2015) Resour Policy 46:272
20. Binnemans K, Jones PT, Blanpain B, Van Gerven T, Yang Y, Walton A, Buchert M (2013) J Clean Prod 51:1
21. Golev A, Scott M, Erskine PD, Ali SH, Ballantyne GR (2014) Resour Policy 41:52
22. Pinkham M (1993) Am Met Mark 101:8
23. Mouza AA, Peolides CA, Paras SV (1995) Resour Conserv Recycl 15:95
24. Hageluken C (2006) METALL-BERLIN 60:31
25. Steel MCF (1991) Stud Surf Sci Catal 71:105
26. Rosso JP (1992) Chem Eng Prog 88:66
27. Gambogi J (2013)
28. Haque N, Hughes A, Lim S, Vernon C (2014) Resources 3:614
29. Nassar NT (2015) Global stocks and flows, losses, and recoveries of platinum-group elements. Yale University
30. Dong H, Zhao J, Chen J, Wu Y, Li B (2015) Int J Miner Process 145:108
31. Bertau M, Froehlich P, Brett B, Lorenz T, Martin G (2016) Angew Chem Int Ed
32. Matthey J (2013) Platin Today
33. Peng Z, Li Z, Lin X, Tang H, Ye L, Ma Y, Rao M, Zhang Y, Li G, Jiang T (2017) JOM 69:1553
34. Ghodrat M, Rhamdhani MA, Sharafi P, Samali B (2017) Metall Mater Trans E
35. Ivanović SZ, Trujuć VK, Gorgievski MD, Mišić LD, Božić DS (2011) 15th international research/expert conference, p 701
36. Fornalczyk A, Saternus M (2013) Metabk 52:219
37. Jha MK, Lee J, Kim M, Jeong J, Kim B-S, Kumar V (2013) Hydrometallurgy 133:23
38. Ghodrat M, Rhamdhani MA, Khaliq A, Brooks G, Samali B, Mater J (2017) Cycles Waste Manag 1
39. Ghodrat M, Rhamdhani MA, Brooks G, Masood S, Corder G (2016) J Clean Prod

References

[illegible]

Leaching Recovery of Silver from Used Radiographic Films

A. A. Adeleke, A. N. Adebayo, B. O. Ibitoye and K. E. Oluwabunmi

Abstract A preliminary leaching recovery of silver metal from waste radiographic films with potassium hydroxide (KOH) in a water bath was investigated. The rinsed and cleansed films were sized into squares before being dried in an oven at 20 °C for 30 min. Silver oxide was precipitated with potassium hydroxide at varying concentrations at 90 °C in a water bath by agitated leaching. The leach solutions were filtered and the filtrates obtained analyzed for silver with Atomic Absorption Spectrophotometer. The results obtained showed that the leach liquors assayed 648.27, 984.86 and 1017.82 mg/L for KOH solutions of molar concentrations 0.02, 0.5 and 1.0 M, respectively. The silver oxide precipitate obtained after filtration was also thermally decomposed at 450 °C. Future research will include analysis for the silver content of the residue resulting from thermal decomposition. The study showed that silver metal in waste radiographic films can be leached with potassium hydroxide in a water bath at 90 °C and thus further confirm waste radiographic films as a secondary resource for silver recovery.

Keywords Radiographic films · Leaching · Potassium hydroxide Filtrate · Precipitation

---

A. A. Adeleke (✉) · A. N. Adebayo
Department of Materials Science and Engineering, Obafemi Awolowo University, Ile-Ife, Nigeria
e-mail: adeadeleke@oauife.edu.ng

A. N. Adebayo
e-mail: adebayoadeniyi35@gmail.com

B. O. Ibitoye
Department of Radiology, Obafemi Awolowo University, Ile-Ife, Nigeria
e-mail: bobitoye@yahoo.com

K. E. Oluwabunmi
Prototype Engineering Development Institute (PEDI), Ilesa, Nigeria
e-mail: kayoxemak@yahoo.com

G. Lambotte et al. (eds.), *Materials Processing Fundamentals 2018*,
The Minerals, Metals & Materials Series,
https://doi.org/10.1007/978-3-319-72131-6_14

Introduction

Silver, with a price of about $500 per kg is one of the precious and noble metals. It is used in large quantities for many purposes, particularly in the photographic industry due to its photosensitive properties. Silver's sensitivity to light has resulted in its wide usage in most photographic and x-ray materials [1]. The scarcity of silver and its immense usefulness for both industrial and domestic purposes make the development of a novel secondary route for its production an urgent need. It has been noted that 25% of the world's silver requirements are obtained by recycling and that 75% of the recycled products originate from photographic waste [2]. In comparison to other films, X-ray photographic films contain an appreciable percentage of silver ranging from 0.7 to 2.0% to enhance sensitivity to transmitted X-rays. The silver in radiographic film is in the form of silver halides and temperatures higher than 22 °C increases the tendency for the reduction of silver into its grains [2–5].

Although silver metal is known to be almost hazard free for both plant and animal lives, its soluble salts and the emissions during its pyro-metallurgical recovery pose a great risk to the ecosystem. Silver compounds can be slowly absorbed by body tissues, with the consequent bluish or blackish skin pigmentation (argyria). In view of this, enzymatic methods using microbial proteases and leaching have been explored in recent years as alternatives to the burning and oxidation methods of silver recovery from photographic or X-ray films. However, the use of microorganisms in bio-leaching was reported to be relatively slow [6–8].

The hydrometallurgical methods reported for the recovery of silver from films generally involved the stripping of silver from the film by leaching and the subsequent recovery of the stripped silver by smelting or electrolysis. In this research, the recycling of silver metal from used radiographic films using potassium hydroxide as the leaching agent at varying concentrations was investigated.

Materials and Methods

Materials

Sample Preparation

Twenty X-ray photographic films were collected from the Radiology Department, Obafemi Awolowo University Teaching Hospital, Ile Ife, Nigeria. The used photographic waste films were washed with distilled water and wiped with cotton wool impregnated with ethanol. The films were oven dried at 20 °C for 30 min to ensure silver remains in its combined form as halide and it was then sized into 2 cm by 2 cm square pieces [3].

Methods

Leaching of the Films in Water Bath

A 50 g mass of the waste radiographic film was placed in 250 ml, 0.02 M potassium hydroxide (KOH) solution contained in a flask in the water bath at 90 °C, marked 250 ml volume, but with higher volume capacity to accommodate the sample and the leach solution for leaching based on stoichiometric calculations. The films were leached by potassium hydroxide in the water bath, with intermittent stirring about 5 min interval, until the reaction with the photographic films were observed to be completed. The films were visually examined to determine the extent of leaching and weighed after the leach stripping. During the leaching, a gelatine-silver layer, the black colloidal solution known as silver oxide was observed in the leach liquor. The time in minutes observed for the complete stripping of the films was noted. The resulting leach solution containing colloidal metallic silver precipitates was filtered into a 250 ml Erlenmeyer flask using a 125 mm diameter Whatman filter paper. The procedure described was repeated but at 0.05 and 1 M concentrations of potassium hydroxide. It has been reported that the reactions to produce the insoluble silver oxide takes place in two steps as follows [9, 10]:

$$\mathrm{KOH}(aq) + \mathrm{AgI}(s) = \mathrm{AgOH}(s) + \mathrm{KI}(aq) \tag{1}$$

The silver hydroxide solid decomposes spontaneously to form silver oxide by the reaction:

$$2\mathrm{AgOH}(s) = \mathrm{Ag_2O}(s) + \mathrm{H_2O}(l) \tag{2}$$

The filtered solution was decanted and the residue washed with distilled water, dried and weighed. A comparison leaching test with 150 ml of 1 M nitric acid instead of potassium hydroxide was also carried out for 30 min. Evolution of the choking, poisonous nitrogen monoxide gas occurred and potassium hydroxide was then added. The reactions occurred as follows:

$$3\mathrm{Ag}(s) + 4\mathrm{HNO_3}(aq) = 2\mathrm{AgNO_3}(aq)2\mathrm{H_2O}(l) + \mathrm{NO}(g) \tag{3}$$

$$2\mathrm{AgNO_3}(aq) + 2\mathrm{KOH}(aq) = \mathrm{Ag_2O}(s) + 2\mathrm{KNO_3}(aq) + \mathrm{H_2O}(l) \tag{4}$$

Thermal Decomposition

The filter paper with the silver oxide precipitate was then placed in a silica crucible and thermally decomposed to silver metal residue in the a muffle furnace at 450 °C.

Visual Observation

The leach stripping solution was visually observed in the Erlenmeyer flask to determine the colour change due to the leaching reactions. After the thermal decomposition, the residue obtained was also visually examined.

Atomic Absorption Spectrophotometer

The sample was analyzed with PG990 Atomic Absorption Spectrometer (AAS). The leach solution sample was flame atomized using air-acetylene flame. The light with characteristic emission spectrum of silver was emitted (under a high voltage applied to the single element (silver) hollow cathode lamp). The light was focused into the flame atomized sample causing emission of photons that formed an atomic spectral line. Based on the quantity of energy input into the flame and the quantity of energy emitted, the concentration of silver was determined [11, 12]. Figure 1 presents a schematic diagram of a typical Perkin Elmer's AAS experimental set up.

Results and Discussion

After the leaching, it was observed that there was a change in the colour of the leach stripping solution and the radiographic films. The films were initially opaque but became increasingly transparent depending on the degree of leaching. The colour changes observed in the leach solutions were noted to differ for nitric acid and potassium hydroxide. The stripping solution appeared blackish for potassium hydroxide and bluish for nitric acid. The comparison tests between the use of potassium hydroxide and nitric acid indicated that potassium hydroxide was a

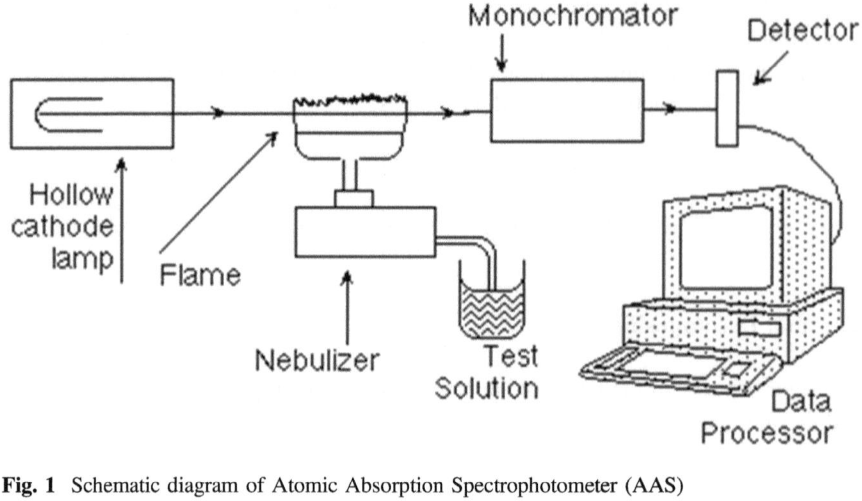

Fig. 1 Schematic diagram of Atomic Absorption Spectrophotometer (AAS)

preferable leaching reagent because nitric acid produced the poisonous nitrogen monoxide gas. The change in colour observed thus strongly suggests that the addition of leaching reagent caused a chemical change in the radiographic films. It was further observed that a grey residue on a black substance was formed in the silica crucible after thermal decomposition.

Figure 2 shows the effects of reagent concentration on the leach contact time. It was noted that the time required for the completion of the reaction of the leachant with the films decreased from 65 min at 0.02 M to 30 min at 1 M concentration of the potassium hydroxide leaching reagent. The rate of chemical reaction generally depends on factors such as concentration of reagents, particle size and temperature [13, 14].

In Fig. 3, the variation in the concentrations of dissolved silver in the potassium hydroxide leach liquor at various concentrations is shown. The results showed that the concentrations of silver in the leach solution increased from 648.02 mg/L at 0.02 M potassium hydroxide leachant concentration to 1017.82 at 1 M concentration, translating to about 36.3% increase in silver recovery. It has been reported that the leaching potency of a chemical reagent generally increases with increasing leachant concentration up to an optimum concentration level [10]. The results obtained suggest that higher concentrations of potassium hydroxide leachant will produce higher concentrations of silver in the leach liquor and an optimal

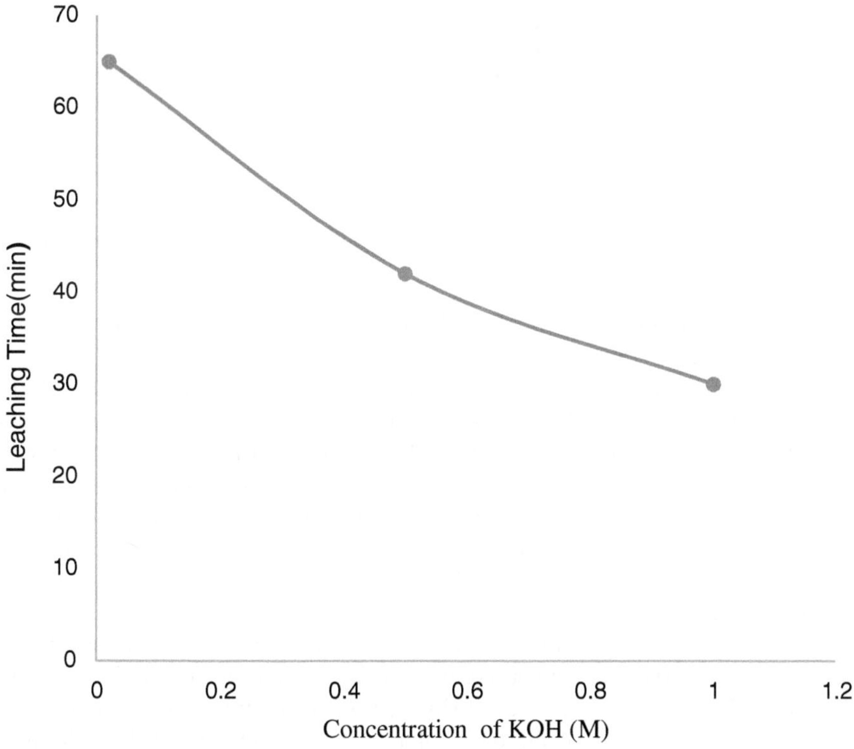

Fig. 2 Variation of leaching time with KOH leachant concentrations

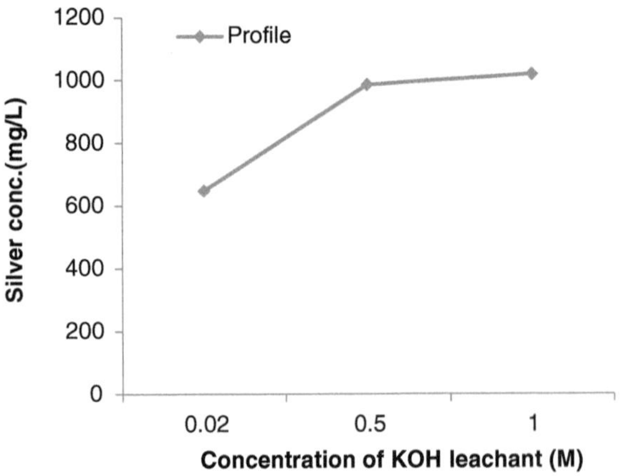

Fig. 3 Variation of silver concentration in leach liquor with KOH concentrations

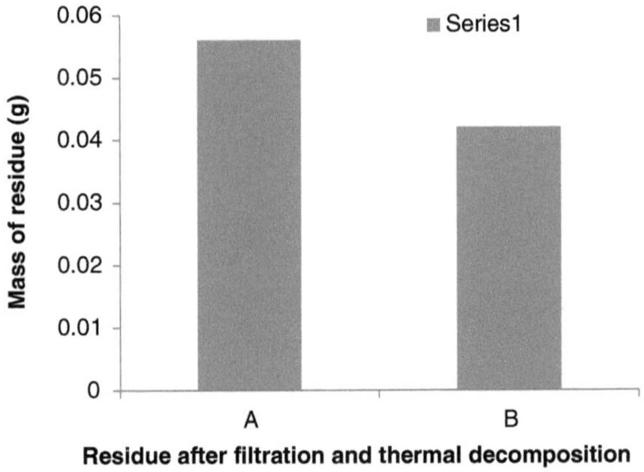

Fig. 4 Weights of residues after filtration and thermal decomposition

combination of leaching time and leachant concentration for maximum economic return has to be determined.

The comparison test with nitric acid that directly leached the silver metal into solution as silver nitrate indicated the evolution of the poisonous nitric mono oxide gas which was not involved in reactions 1 and 2 for KOH. The use of potassium hydroxide as a leaching agent for radiographic films instead of nitric acid will thus eliminate the continuous emission of nitrogen mono oxide (NO), a poisonous gas and the production of silver nitrate solution which when slowly absorbed into the body system to a concentration level that exceeds 2 g will result in a skin condition called argyria [15].

Figure 4 shows the weights of residues after filtration and thermal decomposition. It was observed that the mass of the silver oxide precipitate (A) decreased by 25% and left behind a grey residue of silver (B) on a dark impurity. Future research will include the analysis of the dark coloured precipitates obtained on the filter paper and the grey coloured residue after the thermal decomposition.

Conclusions

Potassium hydroxide has been successfully used to leach silver from waste radiographic films in a water bath at 90 °C. The comparison leaching test with nitric acid was also successful but led to the evolution of the poisonous nitrogen mono oxide gas. Hence, this preliminary research has further shown potassium hydroxide as an environmentally benign leachant for the recovery of silver from radiographic films secondary resource by the simple water bath leaching.

References

1. Mendoza CS, Kanata S (1996) Silver extraction for pollution Control of photographic fixing solution with tetramethytturam disulphide. Bull Chem Soc Jpn 69:3499–3504
2. NDT Resource Centre (2017) Radiographic Film https://www.nde-ed.org/EducationResources/.../Radiography/.../radiographicfilm.htm. Accessed 28 Aug 2017
3. Thayan K (2014) The physics of radiology and imaging. JP Medical Ltd. https://books.google.com.ng/books?isbn=9351521710. Accessed 28 Aug 2017
4. Lakshmi BKM, Hemalatha KPJ (2016) Eco friendly recovery of silver from used X-ray films by alkaline protease of *Bacillus cereus* strain S8. Front Environ Microbiol 2(6):45–48
5. Aktas S, Morcali MH, Yucel O (2010) Silver recovery from waste radiographic films by cementation and reduction. Can Metall Q 49(2):147–154
6. Syed S, Sharma LM, Syed AA (2002) Clean technology for the recovery of silver from processed radiographic films. Hydrometallurgy 63(3):277–280
7. Ingale S, Rele M, Srinivasan M (2002) Alkaline protease production by basidiobolus: effect of 'darmform' morphogenesis and cultural conditions on enzyme production and preliminary enzyme characterization. World J Microbiol Biotechnol 18:403–408
8. Nakiboglu N, Toscali D, Yasa I, Turk J (2001) A novel silver recovery method from waste photographic films with NaOH stripping chemistry. 25:349–353
9. Kirtley M (2017) What causes silver hydroxide to precipitate? https://www.quora.com/What-causes-silver-hydroxide-to-precipitate. Accessed 15 Sept 2017
10. SDN Forum (2017) Solving for pH. https://forums.studentdoctor.net/threads/solving-for-ph-when-to-use-ph-equation-vs-ksp.1057633/. Accessed 15 Sept 2017
11. Weltz B (1985) Atomic absorption spectrophotometry. VCH-Verlag-Ges
12. AZO Materials Analytical Services (2008) Atomic absorption spectroscopy (AAS)-determination of chemical composition of metals and ceramics by LSM analytical services. https://www.azom.com/article.aspx?ArticleID=4222. Accessed 2 Oct 2017
13. Gosh A, Ray HS (1991) Principles of extractive metallurgy. Wiley, New York
14. Upadhyaya GS, Dube RK (1977) Problems in metallurgical thermodynamics and kinetics. Pergamon Press, New York
15. Siverperoxide.com (2017) Exposure controls containment Level 4. http://www.silverperoxide.com/Documents/Danger%20of%20silver%20nitrate.pdf. Accessed 28 Aug 2017

The Study of Copper Leaching from Conichalcite and Chalcopyrite Using Alternative Lixiviants

Junmo Ahn, Isabel F. Barton, Doyun Shin and Jaeheon Lee

Abstract This project investigated alternative lixiviants for leaching copper from chalcopyrite and conichalcite, two highly refractory Cu ore minerals. Chalcopyrite ($CuFeS_2$) is the most abundant of the copper sulfide minerals; conichalcite ($CaCuAsO_4OH$) is less well known but contributes to supergene Cu inventory in high-As deposits. Neither leaches well in sulfuric acid. Sulfurous acid, glycine, methanesulfonic acid (MSA), and thiourea (for conichalcite leaching test) were tested as lixiviants for chalcopyrite- and conichalcite-bearing ores with ferric sulfate or hydrogen peroxide as oxidant. The highest Cu extraction from chalcopyrite was obtained by MSA (47% Cu recovery in 30 g/L of MSA with 5 g/L Fe^{3+} at 75 °C within 96 h; nearly 100% Cu extraction in 30 g/L MSA with 3% hydrogen peroxide at 75 °C within 72 h). Glycine recovered 20.7% of copper from chalcopyrite at room temperature within 96 h. All these compare favorably to the results of sulfuric acid leaching of chalcopyrite (1.5% recovery at 96 h). Conichalcite proved less refractory to sulfuric acid (46–71% extraction in 10 g/L H_2SO_4 with 3 g/L Fe^{3+} in 24 h) and leached almost as well in MSA (45–67% extraction in 20 g/L MSA with 3 g/L Fe^{3+} in 24 h), but glycine, thiourea, and sulfurous acid did not effectively leach conichalcite.

Keywords Copper · Chalcopyrite · Conichalcite · Alternative lixiviants Sulfurous acid · Methanesulfonic acid · Glycine

J. Ahn · I. F. Barton · D. Shin · J. Lee (✉)
Department of Mining and Geological Engineering, University of Arizona, Tucson, AZ, USA
e-mail: jaeheon@email.arizona.edu

I. F. Barton
Lowell Institute for Mineral Resources, University of Arizona, Tucson, AZ, USA

G. Lambotte et al. (eds.), *Materials Processing Fundamentals 2018*,
The Minerals, Metals & Materials Series,
https://doi.org/10.1007/978-3-319-72131-6_15

Introduction

Chalcopyrite ($CuFeS_2$) accounts for about 70% of copper ores on the Earth [1], but is problematic to leach with conventional sulfuric acid/ferric system. Elemental sulfur and jarosite precipitation inhibit copper leaching by passivating the chalcopyrite surface [2]. Conichalcite [$CaCuAsO_4OH$) forms in the oxidized zones of some copper ore bodies [3]. Although it is often mistaken for malachite and therefore not reported, it contributes to Cu inventory in the Carlin trend, Cloncurry, Broken Hill, Pampa Larga, Kolwezi, Laurion, and Morenci, among others (mindat.org; R. Stegen, pers. comm. to IB, 2017). Conichalcite is not well studied in copper hydrometallurgy, but initial tests indicated it is refractory. Therefore, it was included in this study of alternative lixiviants for recovering copper. The lixiviants tested here are sulfuric acid, methanesulfonic acid (MSA), sulfurous acid, and glycine for chalcopyrite, and thiourea is added for conichalcite leaching.

Sulfurous acid (H_2SO_3) has a lower redox potential than sulfuric acid, which may alleviate passivation on the surface of chalcopyrite. Several studies have examined sulfurous acid as an leaching agent to leach copper from various solid samples such as lead blast furnace slag [4], cuprite [5], chalcocite [6], and smelter slags [7], but none has been carried out on chalcopyrite. Methanesulfonic acid (MSA; CH_3SO_3H) is a non-toxic organic acid with high metal salt solubility and stability compared to sulfuric acid [8]. Although relatively new to hydrometallurgy and therefore poorly studied, it has been found to leach copper from malachite (about 92.4% of copper extraction by 0.36 mol/L (35 g/L) MSA at 65 °C after 60 min) [9]. Glycine ($C_2H_5NO_2$) is one of the simplest amino acids and can leach copper, gold and silver at alkaline condition by forming metal-amino acid complexes [10, 11]. Glycine was tested as a lixiviant for a gravity concentrate of covellite (Cu extraction about 19%), bornite (92%), chalcocite (100%), and cuprite (100%) by 0.3 M glycine with 1% H_2O_2 at pH 11 for 48 h at room temperature [10].

The objective of this study is (1) to compare sulfurous acid, MSA, and glycine (±thiourea) as lixiviants for copper leaching in chalcopyrite and conichalcite with conventional sulfuric acid/ferric leaching, (2) to determine optimal leaching conditions for chalcopyrite leaching and (3) to investigate the best performing lixiviant for copper leaching from conichalcite containing ore by comparing Cu extraction rates and analyzing residues.

Materials and Methods

Sample Preparation

Chalcopyrite concentrate (P80 200 mesh) was obtained from Freeport-McMoRan, Inc.'s Sierrita mine (Arizona, USA). The result of elemental composition analysis showed that the concentrate was 28.8% Cu, 26.8% Fe and 31.5% S. About 13% of the sample is gangue. XRD showed that the 87% of ore material in the sample consists of 95% chalcopyrite, 3% pyrite, and 2% cuprite. Five different conichalcite-containing ores were collected from the supergene zone of the Mike deposit (Carlin trend, Nevada, USA) and ground to 10 and 200 mesh. QEMSCAN (Quantitative Evaluation of Mineralogy by Scanning Electron Microscopy) mineralogical analysis showed that the conichalcite-containing ores contain from 0.01–0.66% chrysocolla, 0.2–0.3% copper sulfide, 0.02–0.11% malachite, 0.03–1.36% conichalcite and 0.01–0.33% other copper minerals. Gangue minerals are quartz, sericite, chlorite, alunite, smectite, kaolinite, potassium feldspar, iron oxides and jarosite.

Leaching Tests

Chalcopyrite Leaching Tests

Sulfurous acid, MSA, and glycine were used to leach copper from chalcopyrite and compared to sulfuric acid as a baseline. The test was carried out by varying the concentration of lixiviants from 5 to 50 g/L with 5 g/L ferric ion to compare the extraction capacity of the lixiviants. The concentration of MSA was compared with a half concentration of sulfuric and sulfurous acid because MSA is dissociated into one proton and one methanesulfonate anion. Then, the concentration of lixiviant which showed the best copper extraction ability was set from 10 to 50 g/L to optimize the concentration of lixiviant. After the optimization of lixiviant concentration, ferric concentration was set from 1 to 10 g/L to find the effect of ferric on chalcopyrite leaching for 96 h. Hydrogen peroxide (from 1 to 15% (v/v) and further addition at the same amount every 24 h) was also used as an alternative lixiviant for leaching test at 75 °C for 96 h. All acid beaker leaching tests were carried out at 75 °C. Glycine (1 M) and sulfuric acid (10 g/L) leaching were performed via bottle roll tests with 3% hydrogen peroxide (w/v) and 5 g/L ferric, respectively, at a pulp density of 6.25%. For glycine, pH was maintained at or above 11 by periodic addition of 50% NaOH solution. Bottle roll tests were carried out at room temperature for 24 h with kinetic samples collected at 2, 6, and 24 h with pH and Eh values.

Conichalcite Leaching Tests

Sulfuric acid, sulfurous acid, MSA, thiourea and glycine were used to leach copper from the conichalcite-containing ore. Bottle roll tests were performed with 40% pulp density at room temperature for 24 h. The concentration of leaching agents and oxidants is shown in Table 1. Eh and pH were monitored at 0, 2, 6, and every

Table 1 The condition of lixiviant and oxidant for bottle roll leaching test of conicalchite containing ore leaching with 40% pulp density at 25 °C for 24 h

Lixiviant	Concentration	Oxidant	Concentration	Controlled pH
Sulfuric acid	10 g/L	Fe^{3+}	5 g/L	<1.5
Sulfurous acid	10 g/L	Fe^{3+}	5 g/L	<1.5
MSA	20 g/L	Fe^{3+}	5 g/L	<1.5
Thiourea[a]	10 g/L	Fe^{3+}	3 g/L	<1.5
Glycine	0.5 M	H_2O_2	3%	>11[b]

[a]1 g/L of sodium metabisulfite was also added for the thiourea leaching tests
[b]pH for glycine tests was maintained above 11 by 50% (w/v) sodium hydroxide

24 h. Leachate samples were collected at 2, 6 and 24 h and filtered with 0.45 μm syringe filters. The solid residue was collected after 24 h for further analysis.

Analytical Methods

Copper concentration in the liquid samples was analyzed on a single-lamp Perkin-Elmer Atomic Absorption (AA) spectrophotometer. The solid residue of conichalcite was analyzed by QEMSCAN on an FEI Quanta 400 with a beam current of 5 nA and accelerating voltage of 25 kV. Quantitative mineralogy was based on 5000 particle analyses at a field spacing of 300 microns plus a series of line scans with 1800 microns field size and 10-point spacing. Both analyses were done with a 5-micron pixel size.

Results and Discussion

Chalcopyrite Leaching by Various Lixiviants and Oxidants

The comparison of the leaching capacity of sulfuric acid, sulfurous acid and MSA is shown in Fig. 1. MSA extracted more copper from chalcopyrite than sulfurous and sulfuric acid. The copper recovery (25%) was higher at 20 g/L of MSA with 5 g/L of ferric at 75 °C than among other lixiviants such as 10 g/L sulfuric acid and 10 g/L sulfurous acid with 5 g/L ferric which recovered 18 and 16% of copper, respectively. As shown in Fig. 2, copper extraction increased with increasing concentrations of MSA from 10 g/L to 30 g/L and then decreased. The maximum copper extraction for 24 h at 75 °C was about 27% at 30 g/L MSA and 5 g/L ferric. The effect of acid concentration on copper extraction was more significant in sulfurous acid than in MSA.

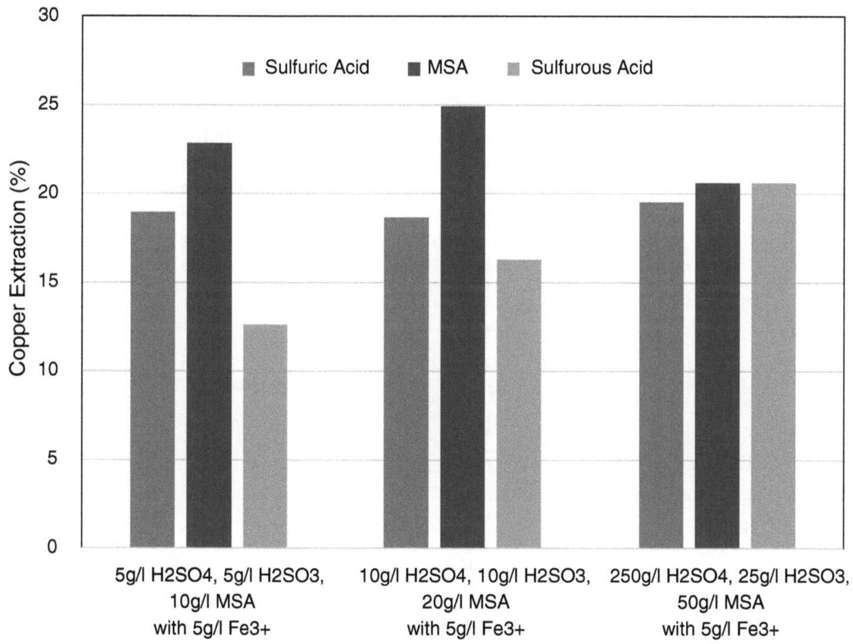

Fig. 1 Copper extractions using various concentrations of sulfuric, sulfurous, and methane sulfonic acids with different oxidants and their concentrations

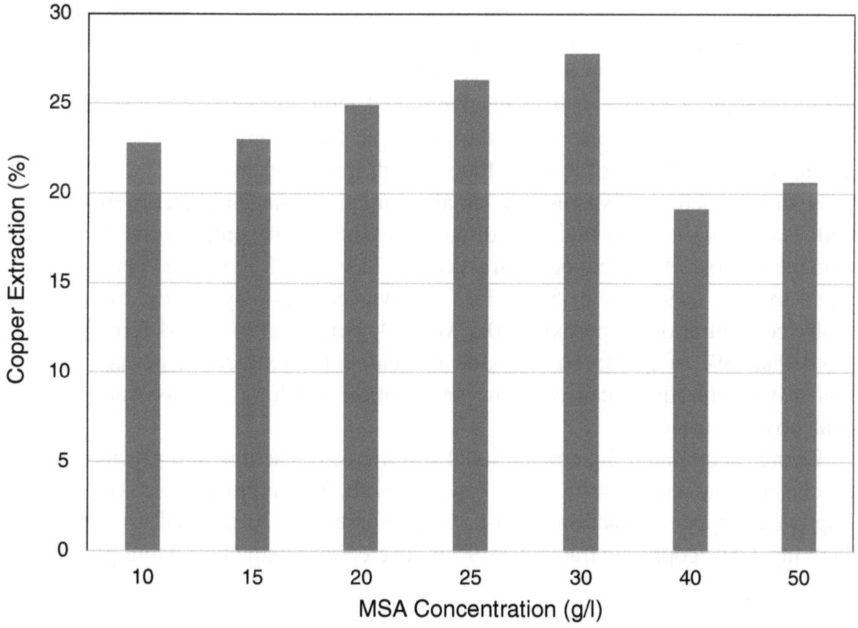

Fig. 2 Effects of MSA concentration on copper extraction for 24 h leaching at 75 °C

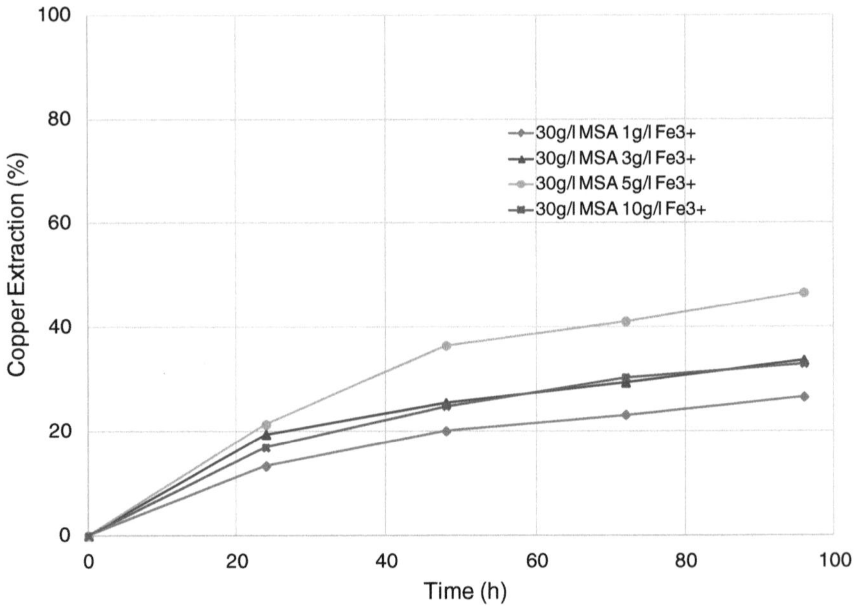

Fig. 3 Effects of ferric concentration on copper extraction with 30 g/L MSA for 96 h leaching at 75 °C

The effect of ferric ion concentration on copper extraction with 30 g/L MSA was investigated and showed in Fig. 3. A condition of 5 g/L ferric ion showed about 47% of copper extraction in 96 h, whereas with 3 and 10 g/L ferric ion showed copper extractions about 34 and 33% respectively, and 1 g/L of ferric extracted about 27% of copper. The lower copper extraction at 10 g/L ferric compared to 5 g/L might be due to the precipitation of ferric sulfate at high iron concentration.

Hydrogen peroxide was used as an alternative oxidant for chalcopyrite leaching with MSA. As shown in Fig. 4, hydrogen peroxide clearly outperformed Fe^{3+} as an oxidant (almost 100% copper extraction in 72 h at 3% hydrogen peroxide compared to 47% at 96 h with 5 g/L Fe^{3+}). However, this may be due to the higher molar concentration of peroxide (0.5 M H_2O_2 compared to 0.1 M ferric). Adding more than 3% of hydrogen peroxide decreased the extraction, perhaps because higher concentrations causes iron precipitation resulting in passivation of the chalcopyrite surface.

Figure 5 compared the conventional sulfuric acid leaching and glycine leaching of chalcopyrite concentrate. 1 M Glycine with 3% hydrogen peroxide showed improved copper extraction of 14.3% compared to 10 g/L sulfuric acid only showing 1.3% copper extraction in 96 h. The copper extraction using glycine was

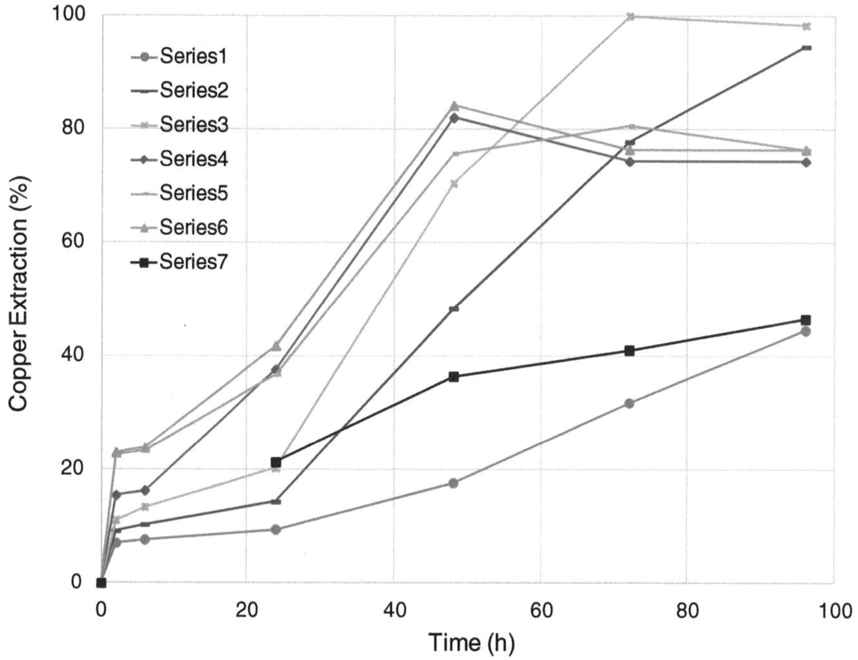

Fig. 4 Effects of oxidants (hydrogen peroxide and ferric ion) concentration on copper extraction with 30 g/L MSA for 96 h leaching at 75 °C

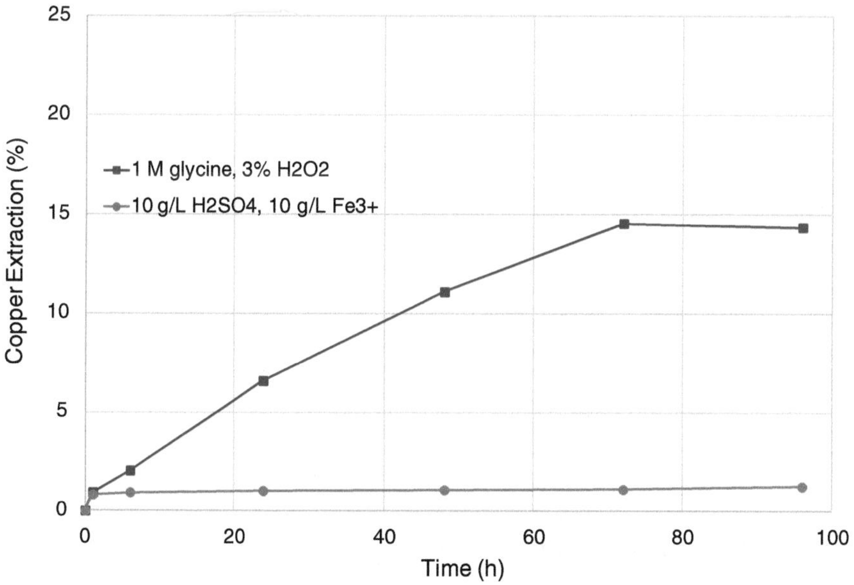

Fig. 5 Comparison of copper extraction using 10 g/L sulfuric acid with 10 g/L ferric ion and 1 M glycine with 3% H_2O_2

still lower than that of MSA. This indicates that glycine can be used to leach copper from chalcopyrite to some extent, although this remains to be verified through more comprehensive study.

Conichalcite Leaching by Various Lixiviants and Oxidants

The results of 24 h leaching by sulfuric acid, sulfurous acid, glycine, MSA, and thiourea are shown in Table 2. All three acids tests extracted more copper than glycine and thiourea. Optimum conditions varied from sample to sample, but sulfuric acid and MSA were consistently the most effective lixiviants. For samples 1, 2, and 5, 10 g/L sulfuric acid with 5 g/L ferric yielded maximum copper recovery (70.8% at 10 mesh and 44.8% at 200 mesh for sample 1, 42.8% at 10 mesh and 27.0% at 200 mesh for sample 2, 31.4% at 10 mesh and 37.1% at 200 mesh for sample 5). For samples 3 and 4, 20 g/L MSA and 5 g/L ferric was most effective (69.8% at 10 mesh and 78.1% at 200 mesh, 68.4% at 10 mesh and 88.9% at 200 mesh respectively). Sulfurous acid recovered less copper than sulfuric acid or MSA for all samples, but more than glycine or thiourea. No consistent relationship was observed between particle size and copper leaching.

Table 3 shows a typical example of the mineralogical changes from the head to the leach residues. After 24 h of leaching, no copper minerals were detected in the residues from sulfuric acid and MSA. Sulfurous acid had about 0.2% residual conichalcite, and the glycine and thiourea residues had about 1.6 and 0.95% respectively. These results indicate that sulfuric acid and MSA are the best lixiviants for conichalcite and conichalcite-bearing ores.

Table 2 Copper extraction (%) from conichalcite after 24 h' leaching at room temperature

Particle size (mesh)	Lixiviant	Oxidant concentration	Cu extraction (%) at 24 h				
			Sample 1	Sample 2	Sample 3	Sample 4	Sample 5
10	10 g/L H_2SO_4	5 g/L Fe^{3+}	71	43	61	72	31
200	10 g/L H_2SO_4	5 g/L Fe^{3+}	45	27	52	65	37
10	10 g/L H_2SO_3	5 g/L Fe^{3+}	38	28	53	67	20
200	10 g/L H_2SO_3	5 g/L Fe^{3+}	25	22	53	60	19
10	0.5 M Glycine	3% H_2O_2	3.4	6.9	20	20	20
200	0.5 M Glycine	3% H_2O_2	3.0	8.5	20	19	10
10	20 g/L MSA	5 g/L Fe^{3+}	46	29	70	68	25
200	20 g/L MSA	5 g/L Fe^{3+}	67	27	78	89	30
10	10 g/L Thiourea	3 g/L Fe^{3+}	2.4	8.7	15	33	1.7
200	10 g/L Thiourea	3 g/L Fe^{3+}	3.3	11	4.7	26	2.6

Table 3 Copper mineral %s in 10-mesh samples of head and leach residues[a]

Minerals (% of sample)	Head	Sulfuric acid residue	MSA residue	Sulfurous acid residue	Glycine residue	Thiourea residue
Pyrite (FeS_2)	0.40	0.23	0.21	0.24	0.19	0.23
Chrysocolla (Cu-silicate, various formulas)	0.07	–	–	0.01	0.07	0.04
Malachite ($Cu_2(CO_3)(OH)_2$)	0.02	–	–	–	0.01	0.01
Conichalcite ($CaCu(AsO_4)(OH)$)	1.36	–	–	0.20	1.60[b]	0.95
Cu in micas and clays (various)	0.04	–	–	0.01	0.05[b]	0.03
Other Cu	0.03	–	–	–	0.02	–

[a]The reader will note that these changes in Cu mineralogy do not match the reported recoveries in Table 2. The amount of Cu measured in the leach liquors reconciles to well within 15% of the difference between chemical assays of head and tail Cu content, but reconciliation with QEMSCAN data was noticeably poorer. This is probably a result of the high fundamental sampling error associated with a 3-g QEMSCAN split. The QEMSCAN data are given here as a guide to the extent of dissolution of different Cu minerals, not as a quantitative evaluation of Cu recovery, for which head-tail and leach liquor chemical assays are much more reliable
[b]The apparent increase in some minerals from head to residue is probably due to QEMSCAN imprecision and some heterogeneity between splits

Conclusion

Sulfurous acid, MSA, and glycine were compared to sulfuric acid as lixiviants for chalcopyrite. MSA showed the highest copper extraction efficiency, reaching 47% extraction with 30 g/L of MSA with 5 g/L ferric at 75 °C and up to 100% extraction with 30 g/L MSA with 3% hydrogen peroxide at 75 °C. Temperature had no effect on recovery up to 45 °C but above that point, recovery increased with increasing leaching temperature. Glycine showed higher Cu extraction (14.3%) than sulfuric acid (1.5%) at room temperature but requires more study. Sulfuric acid, sulfurous acid, MSA, glycine, and thiourea were also tested on conichalcite-bearing ore samples. Although optimum conditions varied with the mineralogy of the samples, sulfuric acid and MSA with ferric consistently recovered more Cu from the conichalcite-containing ores than other lixiviants.

References

1. Córdoba E, Muñoz J, Blázquez M, González F, Ballester A (2008) Leaching of chalcopyrite with ferric ion. Part I: General aspects. Hydrometallurgy 93(3):81–87
2. Klauber C (2008) A critical review of the surface chemistry of acidic ferric sulphate dissolution of chalcopyrite with regards to hindered dissolution. Int J Min Process 86(1):1–17

3. Reddy BJ, Frost RL, Martens WN (2005) Characterization of conichalcite by SEM, FTIR, Raman and electronic reflectance spectroscopy. J Miner Sci 69(2):155–167
4. Sato T, Lawson F (1983) Differential leaching of some lead smelter slags with sulfurous acid and oxygen. Hydrometallurgy 11(3):371–388
5. Youzbashi A, Dixit S (1993) Leaching of Cu_2O with aqueous solution of sulfur dioxide. Metall Mater Trans B 24(4):563–570
6. Zhang W, Singh P, Muir D (2000) SO_2/O_2 as an oxidant in hydrometallurgy. Miner Eng 13 (13):1319–1328
7. Ahmed IB, Gbor PK, Jia CQ (2000) Aqueous sulphur dioxide leaching of Cu, Ni Co, Zn and Fe from smelter slag in absence of oxygen. Can J Chem Eng 78(4):694–703
8. Gernon M (1999) Environmental benefits of methanesulfonic acid. Comparative properties and advantages. Green Chem 1(3):127–140
9. Feng Q, Wen S, Zhao W, Lv C, Bai X (2015) Leaching of copper from malachite with methane-sulfonic acid. Solvent Extr Res Dev 22(2):159–168
10. Oraby E, Eksteen J (2014) The selective leaching of copper from a gold–copper concentrate in glycine solutions. Hydrometallurgy 150:14–19
11. Oraby E, Eksteen J (2015) The leaching of gold, silver and their alloys in alkaline glycine–peroxide solutions and their adsorption on carbon. Hydrometallurgy 152:199–203

Effect of Chloride Ions on the Copper Extraction Using LIX 984N and Acorga M5910

M. C. Ruiz, J. Risso, R. Sanchez and R. Padilla

Abstract Presently, several copper mining plants in Chile are treating by solvent extraction aqueous solutions that contain high concentrations of chloride ions. The effect of chloride ions on the extraction of copper from chloride-sulfate solutions using the commercial extractants LIX 984N and Acorga M5910 was investigated. Copper extraction isotherms were determined at 25 and 35 °C for aqueous solutions of pH 2, containing 6 g/l of Cu(II), 5 g/l of Fe(II) and 2 g/l of Fe(III) with 0, 60 and 110 g/l of chloride ions. The organic solutions used were 26%v/v solutions of LIX 984N or Acorga M5910 in Escaid 110. It was determined that the presence of chloride ions affected negatively the copper extraction equilibria for both extractants, while a temperature increase from 25 to 35 °C produced only a small increment in the copper extraction. On the other hand, a temperature increase affected negatively the stripping equilibria.

Keywords Solvent extraction · Copper extraction · Hydroxyoximes
LIX 984N · Acorga M5910

Introduction

In Chile, heap leaching with sulfuric acid solutions followed by solvent extraction is the preferred technology to treat copper oxide minerals as well as secondary sulfide minerals. Solvent extraction allows obtaining a purified concentrated solution from which copper can be efficiently recovered by electrowinning. In many plants, the leaching solutions contain large quantities of chloride ions originated from the content of the mineral atacamite ($Cu_2(OH)_3Cl$) in the ore, and in some cases, from the sea water used in the sulfuric acid leaching of copper oxide minerals [1–3]. On the other hand, in leaching copper sulfides chloride salts may be added to

M. C. Ruiz (✉) · J. Risso · R. Sanchez · R. Padilla
Department of Metallurgical Engineering, University of Concepcion,
Concepcion, Chile
e-mail: maruiz@udec.cl

G. Lambotte et al. (eds.), *Materials Processing Fundamentals 2018*,
The Minerals, Metals & Materials Series,
https://doi.org/10.1007/978-3-319-72131-6_16

improve the dissolution kinetics of these minerals as an alternative to bioleaching. Therefore, the chloride concentration of the latter solutions can be higher than 100 g/l [3].

The industrial SX plants that are currently treating sulfate-chloride leaching solutions are using hydroxyoxime extractants that have been traditionally used for copper extraction from chloride free sulfate solutions. The more commonly used reagents are the ketoxime 2-hydroxy-5-nonylacetophenone oxime, and the salicyladoxime 5-nonylsalicylaldoxime, or mixtures of them. The basic modification in the solvent extraction circuits when treating chloride containing leaching solutions is the addition of one or more washing stages of the loaded organic solution before stripping, in order to control the transference of chloride ions to the electrowinning electrolyte [1, 4, 5].

The hydroxioxime extractants are able to extract the Cu(II) selectively from the leaching solutions leaving behind most of the impurities. The only other cation that can be extracted to a certain extent is Fe(III). Therefore, copper/iron selectivity was a main concern with the first generation of hydroxyoxime extractants used in industrial applications. It is claimed that the extractants presently used for copper extraction have improved significantly the Cu/Fe selectivity. However, it has been reported that in some cases, the chemically transferred iron to the electrowinning electrolyte can be of a similar magnitude than the iron transferred by entrainment of the leaching solution [6, 7].

The presence of chloride ions changes the ionic speciation in the leaching solutions due to the formation of chloro-complexes of copper and iron cations. This, in turn should affect the equilibrium of the solvent extraction of Cu(II) with hydroxyoxime reagents, and could also affect the extraction of Fe(III). Previous results have shown that the presence of chloride in the aqueous solution has a large negative effect on the copper extraction isotherms of the pure ketoxime, 2-hydroxy-5-nonylacetophenone oxime (LIX 84-IC). The observed effect was smaller for the 5-nonylsalicylaldoxime (LIX 860N-IC) [8].

Considering the above, the objective of the present work was to study the effect of chloride ions on the extraction of copper from chloride-sulfate solutions using two commonly used commercial hydroxyoxime extractants: LIX 984N and Acorga M5910. The organic LIX 984N is a 50/50 v/v mixture of 2-hydroxy-5-nonylacetophenone oxime and 5-nonylsalicylaldoxime, while Acorga M5910 is a modified 5-nonylsalicylaldoxime. The probable effect of chloride ions on copper/iron selectivity was also assessed.

Experimental

The aqueous solutions used in the present study contained 6 g/l of Cu(II), 5 g/l of Fe(II) and 2 g/l of Fe(III), which were added as sulfate salts. Sodium chloride was also added to some solutions to adjust the chloride concentrations to 60 and 110 g/l. In all cases, the pH of the solutions was brought to 2.0 by using sulfuric acid. The organic solutions used were 26% v/v solutions of LIX 984N or Acorga M5910 in Escaid 110.

The experimental work included firstly the determination of the loading capacity of the two hydroxyoxime extractants. Secondly, the extraction isotherms for both extractants at 25 and 35 °C for 0, 60 and 110 g/l of chloride ions in the aqueous phase. Finally, the stripping isotherms at 25 and 35 °C were determined for both loaded organic extractant solutions, using a synthetic spent electrolyte containing 35 g/l of Cu and 180 g/l or 190 g/l of sulfuric acid.

The procedure for the determination of the maximum loading capacity of the hydroxyoxime reagents (g/l of Cu loaded per 1% v/v) consisted of contacting 50 ml of the organic phase (containing 10 to 25% v/v of extractant) with an equal volume of the aqueous phase in a separation funnel. After 5 min of energetic mixing, the two phases were allowed to separate and the aqueous phase was removed through the bottom stopcock. The procedure was repeated by adding fresh aqueous phase until the pH of the aqueous phase did not show any change before and after contact. The complete procedure was carried out at room temperature. The copper loaded into the organic phase was determined by discharging 25 ml sample of the organic by repeated contacts with a 180 g/l sulfuric acid solution in a small separation funnel. The aqueous solutions removed after each contact were mixed and analyzed by atomic absorption spectroscopy (AA).

The determination of the extraction and stripping isotherms was carried out by mixing the organic and aqueous phases until reaching equilibrium in a batch stirred vessel. The reactor was a cylindrical glass reactor with four stainless steel baffles of 1 L capacity. During the experiments, the reactor was immersed in a constant temperature water bath. The two phases were loaded and allowed to reach the desired temperature after which they were stirred at 600 rpm for 10 min. The organic phase (26% v/v of LIX 984N or Acorga M5910), was contacted with the aqueous solutions in various organic/aqueous (O/A) ratios in the range 3/1–1/6 to obtain various equilibrium points. After the specified mixing period the two phases were allowed to separate in the reactor. Once the phases were completely separated, samples of the organic and aqueous phases were withdrawn and filtered to eliminate any entrainments. The aqueous samples were analyzed for copper by AA and the copper loaded into the organic was calculated by difference from the copper concentration before contact. The iron extraction was determined by discharging completely an organic phase sample by repeated contacts with a sulfuric acid solution.

Results

Maximum Loading Capacity

The results of these tests are presented in Fig. 1, which shows the copper loaded into the organic phase for each concentration of the hydroxyoxime reagents in the organic solution.

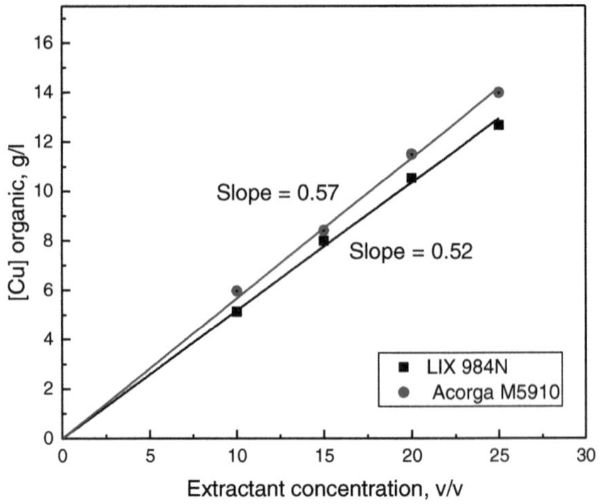

Fig. 1 Determination of maximum loading capacity for LIX 984N and Acorga M5910

The results showed that in both cases the amount of copper charged into the organic phases was directly proportional to the extractant concentration. It has been reported that hydroxyoxime extractants present some degree of dimerization in the organic phase, particularly when using high concentrations of the reagents [9]. However, it is clear from the results in Fig. 1 that dimerization, if occurred, does not affect the maximum loading capacity of these extractants. According to the values of the slopes, the modified salicylaldoxime Acorga M5910 charged slightly more copper per 1% v/v than the mixed hydroxyoxime extractant. Therefore, an organic solution with 26% of Acorga M5910 or LIX 984N will charge 14.8 and 13.5 g/l of copper, respectively. It must be noted that these values of maximum loading capacity would change for a different temperature and composition of the organic phase, particularly the hydrogen ions concentration.

Extraction Isotherms

Extraction isotherms were determined for both solvent extraction reagents using aqueous solutions with chloride concentrations of 0, 60 and 110 g/l. The results obtained for LIX 860N at 25 °C are presented in Fig. 2.

As seen in this figure, the presence of chloride in the aqueous solution has a negative effect on the extraction equilibrium indicated by the shifting of the equilibrium isotherms to the right in the presence of chloride in the aqueous solution. This result is as expected since it was determined previously by the present authors [8] that the presence of chloride ion in the aqueous solution affected negatively the extraction equilibria of the two hydroxyoxime extractants, which are the components of the LIX 984N.

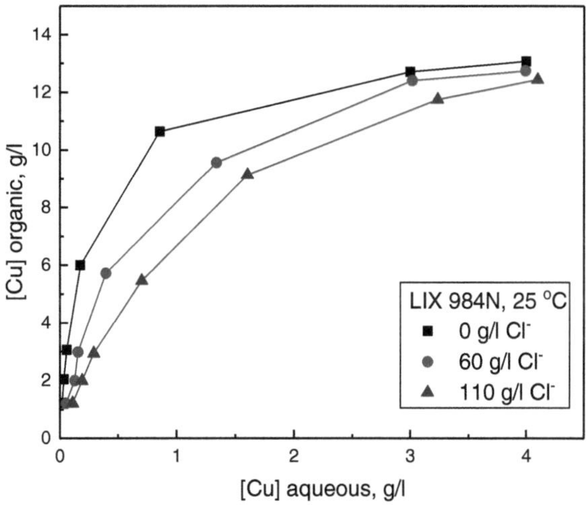

Fig. 2 Extraction isotherms for LIX 984N 26% v/v at 25 °C

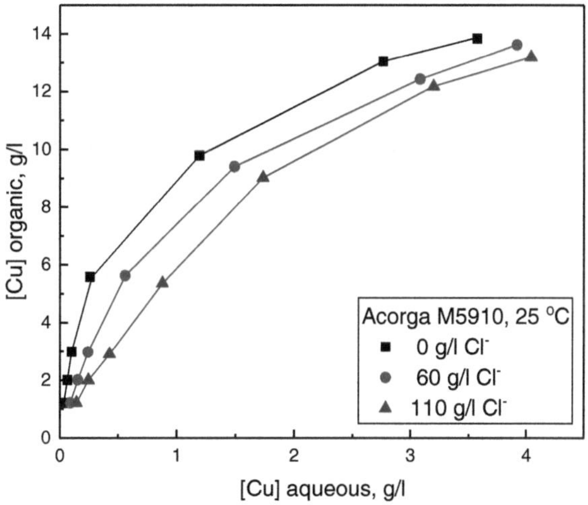

Fig. 3 Extraction isotherms for Acorga M5910, 26% v/v at 25 °C

The extraction isotherms for the modified nonylsalycylaldoxime Acorga M5910 are depicted in Fig. 3.

Again, it can be seen that the extraction equilibria of this extractant is also affected by the presence of chloride ions in the aqueous solution, although the effect is lower than for the LIX 984N, as seen in Fig. 2. This smaller effect could be explained by the content of ketoxime of LIX 984N, which have been shown to be more sensitive to the presence of chloride than the nonyl salicylaldoxime [8]. From Figs. 2 and 3, it is clear that in both cases, the presence of chloride ions in the aqueous solutions will deteriorate the performance of the extractants, leading to an increase of the copper concentration in the raffinate, and a lower copper loading in the organic phase as compared to the pure sulfate solution.

Effect of the Temperature on the Extraction Equilibria

The experimental results showed that a change in temperature from 25 to 35 °C has a small positive effect on the extraction equilibria for both extractants. This is illustrated in Fig. 4 for the aqueous solution containing 110 g/l of chloride ions. In this figure, the dashed lines correspond to 35 °C. It can also be observed in the figure that the extraction isotherms for both extraction reagents were close to one another.

Effect of Temperature on the Stripping Equilibria

Stripping isotherms were also determined for temperatures of 25 and 35 °C and some of the results are presented in Fig. 5. In both cases the organic phase contained 26% v/v of the extractant and it was loaded with 9.6 g/l of Cu for the case of LIX 984N and 11.8 g/l of copper for the case of Acorga M5910. The aqueous phase was a synthetic spent electrolyte with 35 g/l of Cu and 180 g/l or 190 g/l of sulfuric acid.

Since the conditions of the stripping experiments were not the same for both extractants, a direct comparison of the striping isotherms is not possible. However, it is clear that a higher concentration of sulfuric acid in the electrolyte will improve the stripping of copper from the organic phase. In addition, an increase in the temperature from 25 to 35 °C affects negatively the stripping equilibria for both extractants. Therefore, an efficient cooling of the spent electrolyte would be advantageous for the transfer of copper in the striping process.

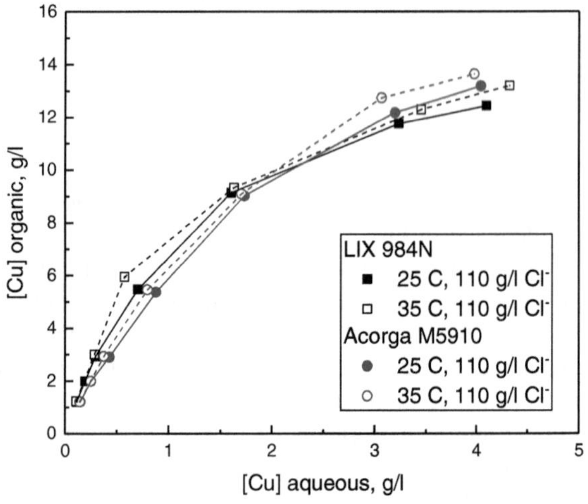

Fig. 4 Effect of temperature on the extraction equilibria for LIX 984N and Acorga M5910. Aqueous phase with 110 g/l of chloride ions

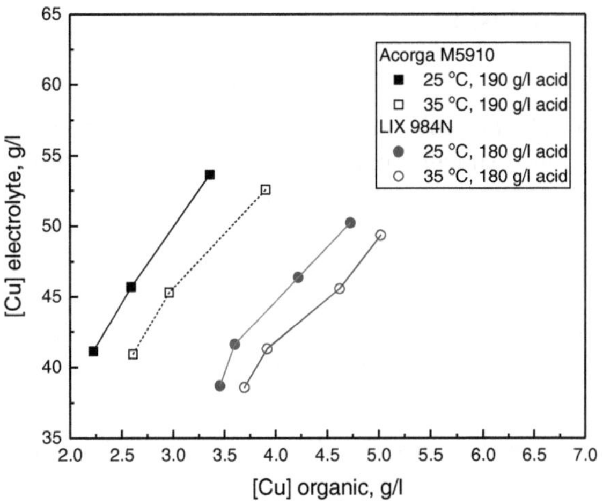

Fig. 5 Stripping isotherm for Acorga M 5910 and LIX 984N at 25 and 35 °C

Copper-Iron Selectivity

The equilibrium concentration of iron in the organic phase was determined for extraction experiments with different O/A ratios and different chloride ions concentrations in the aqueous solutions. Some of the results for the case of LIX 984N are presented in Table 1. The concentration of copper in the organic phase for the same tests are also included in the last column.

From these results we can conclude that the LIX 984N extracted very little Fe (III). This could be due to the large amount of copper loaded into the organic phase in these experiments, which displaced the Fe(III). This displacement of Fe(III) with Cu(II) has been reported previously and it is known as 'crowding' [7]. Nevertheless, the equilibrium iron concentration showed an increment when chloride ions were present in the aqueous phase.

Table 1 Iron extraction at 25 °C for LIX 984N 26% v/v, for various conditions

O/A ratio	$[Cl^-]_{aq}$, g/l	$[Fe]_{org}$, ppm	$[Cu]_{aq}$, g/l
1/4	0	7	12.7
1/6	0	7	13.1
1/4	60	10	12.4
1/6	60	12	12.8
1/4	110	11	11.8
1/6	110	12	13.2

Conclusions

- The presence of chloride ions in the aqueous solution has a negative effect on the extraction equilibria of both Acorga M5910 and LIX 984N.
- An increase in the temperature from 25 to 35 °C has a small positive effect on the extraction equilibria for both hydroxyoxime extractants.
- Increasing the temperature from 25 to 35 °C has a deleterious effect on the stripping equilibria.
- The extraction of Fe is higher when chloride ions are present in the aqueous solution. Nevertheless, the Cu/Fe selectivity is high for LIX 984N.

Acknowledgements The National Fund for Scientific and Technological Development, FONDECYT, of Chile is acknowledged for the financial support of this study through Project No. 1150343.

References

1. Ardiles L, Jolly P (2014) Experiencia industrial con alta concentración de cloruro. Paper presented at the workshop COPPERCLOR 2014, Santiago, Chile, 8–9 May 2014
2. Schmidt P (2013) Usos de agua de mar en plantas hidrometalúrgicas. Paper presented at the 5th international seminar on process hydrometallurgy, Hydroprocess 2013, Santiago, Chile, 10–12 July 2013
3. Schmidt P, Backit A, Jacob J (2013) Cuprochlor®: hydrometallurgical technology for leaching sulfide ores in chloride media. In: Proceedings of Copper 2013, IIMCH, vol IV, pp 43–53
4. Sarangi K et al (2007) Separation of iron(III), copper(II) and zinc(II) from a mixed sulphate-chloride solution using TBP, LIX 84I and cyanex 923. Sep Purif Technol 55:44–49
5. Regel-Rosocka M, Alguacil F (2013) Recent trends in metal extraction. Revista de Metalurgia Madrid 49(4):293–315
6. Simpson J, Navarro P, Alguacil F (1996) Iron(III) extraction by LIX 860 and its influence on copper(II) extraction from sulphuric acid solutions. Hydrometallurgy 42:13–20
7. Hein H (2005) The importance of a wash stage in copper solvent extraction. Paper presented at the workshop Hydrocopper 2005, Santiago, Chile, 24–25 Nov 2005
8. Ruiz MC, Gonzalez I, Salgado J, Padilla R (2017) Extraction of copper from sulfate-chloride solutions by using hydroxyoxime extractants. In: Wang S (ed) Applications of process engineering principles in materials processing, energy and environmental technologies, TMS. Springer, New York, pp 161–168
9. de San Miguel ER et al (1997) Extraction of Cu(II), Fe(III), Ga(III), Ni(II), In(III), Co(II), Zn (II) and Pb(II)with LIX 984 dissolved in n-heptane. Hydrometallurgy 47:19–30

$CaCl_2$–O_2 Roasting of Stibnite and a Complex Copper Concentrate at 500–650 °C

R. Padilla, G. Brito and M. C. Ruiz

Abstract Complex copper sulfide concentrates normally contain As, Sb, Bi, Zn, Pb, as the main impurities. Because of the high levels of these impurities, conventional smelting due to the high risk of environmental pollution cannot treat these concentrates. Oxidizing roasting has been used to remove the unwanted impurities through the gas phase, a process where arsenic is removed effectively but antimony and bismuth removal is very deficient. Because of the poor removal of antimony in oxidizing roasting, an alternative chloridizing roasting of pure stibnite and complex MH copper concentrate (Ministro Hales mine, CODELCO, Chile) has been studied. In roasting in the range of 500–650 °C, temperature and oxygen concentration affected significantly the Sb volatilization in the case of solely Sb_2S_3–$CaCl_2$ sample. At 600 °C and 10% oxygen, antimony volatilization was 60% in 30 min. The volatilization proceeded through intermediate oxides and formation of nonvolatile antimony oxide Sb_6O_{13} as confirmed by XRD spectroscopy. In roasting the MH concentrate at 600 °C and 5% oxygen, arsenic from enargite can be eliminated in less than 20 min most likely as a mixture of arsenic trichloride and trioxide. Oxygen concentration in the gas phase affects not only the rate of weight loss due to volatilization of arsenic, but also the rate of formation of non-volatile compounds.

Keywords Calcium chloride · Roasting · Stibnite · Enargite · Complex-copper concentrate

Introduction

Arsenic and antimony are common impurities in copper concentrates, which are present as the sulfide minerals enargite and stibnite, respectively. Smelters impose heavy penalties on the concentrates when the arsenic content for example exceeds

R. Padilla (✉) · G. Brito · M. C. Ruiz
Department of Metallurgical Engineering, University of Concepcion,
Edmundo Larenas 285, Concepcion, Chile
e-mail: rpadilla@udec.cl

G. Lambotte et al. (eds.), *Materials Processing Fundamentals 2018*,
The Minerals, Metals & Materials Series,
https://doi.org/10.1007/978-3-319-72131-6_17

the permissible limit of about 0.5% [1]. Therefore, concentrates containing high levels of toxic elements must be pretreated before smelting to avoid the risk of environmental pollution. Oxidizing roasting is the classical method used for the pretreatment of dirty copper concentrates to remove the unwanted impurities through the gas phase, a process where arsenic is removed effectively but antimony and bismuth removal is poor at the roasting temperatures (500–700 °C) due to the formation of nonvolatile oxides [2–4]. Other alternatives such as hydrometallurgical methods for the removal of the toxic elements from concentrates have been also proposed [5, 6]. However, these methods have not reached commercial application yet. Thus, presently, oxidizing roasting is still the preferred method for the arsenic elimination.

Thus considering the poor removal of antimony in oxidizing roasting, an alternative chloridizing roasting has been investigated. The aim of this study was the determination of the feasibility of removing arsenic, antimony and bismuth from copper concentrates at the temperature range of 500–650 °C. The upper limit 650 °C was chosen to avoid sintering of the solid particles due to melting of stibnite and tennantite 650 and 657 °C, respectively.

In 1988, Holmstrom [7] conducted diverse experiments using various copper-gold-silver complex concentrates to chloridize As, Sb and Bi by using sodium and calcium chlorides. The experiments included batch tests using up to 10 g of samples and continuous experiments in a rotary furnace using up to 2 kg/h of solids, and smelting tests in an induction furnace. From the entire experimental program, this investigator concluded that As and Sb were removed completely at 500–700 °C, but bismuth removal was better at 600 °C and higher temperatures.

As already indicated, in an industrial oxidizing roasting antimony and bismuth removal is low. Nevertheless, in this paper only the data concerning the chloridizing of antimony and arsenic are discussed.

Experimental

The chloridizing of stibnite and the chloridizing of a complex copper concentrate containing arsenic, antimony and other minor elements were studied by thermogravimetric methods. The experiments were conducted in a thermogravimetric analysis (TGA) setup and in a horizontal tube furnace setup equipped with ancillary equipment including a temperature controller and a gas system to maintain a controlled atmosphere in the reaction tube. The variables studied were temperature, partial pressure of oxygen and concentration of calcium chloride. The gas flow in all the experiments was kept constant at 0.4 L/min.

Table 1 Chemical analysis of complex MH copper concentrate

Element	Cu	Fe	As	Zn	Sb	Pb	Bi
Weight %	33.1	13.30	4.25	0.8	0.23	0.05	<37 ppm

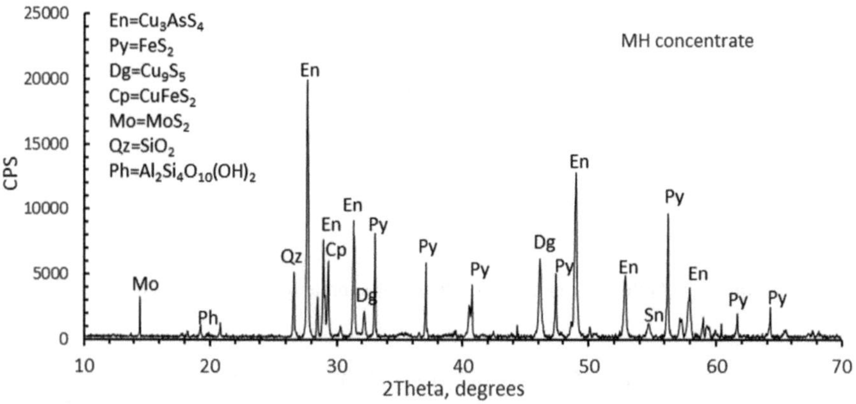

Fig. 1 X-ray diffraction analysis of the complex MH copper sulfide concentrate

Materials

Analytical grade antimony sulfide (Sigma- Aldrich), calcium chloride (Merck), and an industrial complex copper concentrate (MH) from Ministro Hales mine, CODELCO, Chile, were used in the chloridizing experiments. The copper concentrate was used as received and its chemical analysis is shown in Table 1. Additionally, the results of the X-ray diffraction analysis of this concentrate is shown in Fig. 1, which shows diffraction lines for the main minerals enargite (Cu_3AsS_4), pyrite (FeS_2), digenite (Cu_9S_5), and chalcopyrite ($CuFeS_2$).

The diffraction lines for the antimony mineral do not appear in the figure most likely due to the low content of antimony in the concentrate. Considering the chemical analysis, the calculated enargite content in the concentrate was 22.34% and the stibnite content was 0.32%.

Results

The chloridizing reaction of Sb_2S_3 with $CaCl_2$–O_2 was investigated by using the Sb_2S_3 sample to determine the feasibility of chloridizing-volatilization of antimony as a chloride compound. The overall reaction considered is as follows:

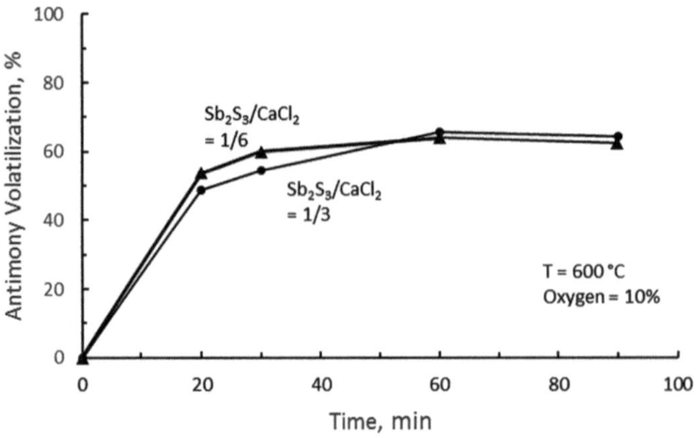

Fig. 2 Antimony volatilization in the chloridizing of stibnite during calcium chloride roasting

$$Sb_2S_3 + 3CaCl_2 + 6O_2(g) = 2SbCl_3(g) + 3CaSO_4 \tag{1}$$

According to this reaction, the stoichiometric $Sb_2S_3/CaCl_2$ molar ratio is equal to 1/3. The results of preliminary experiments obtained in the horizontal tube furnace are shown in Fig. 2. In this figure, the antimony volatilization using stoichiometric $CaCl_2$ is compared to volatilization using 100% excess of $CaCl_2$. It can be seen that the concentration of calcium chloride did not affect the rate of antimony volatilization, thus the use of excess calcium chloride over the stoichiometric amount to enhance the volatilization of antimony may not be necessary. However, even though antimony volatilization is very rapid in the first 30 min, it levels off at around 60% volatilization, indicating that an important fraction of the initial antimony remained in the calcine.

Figure 3 shows the results of XRD analysis of a calcine obtained at 600 °C, 10% oxygen and 30 min of reaction time. This spectrum displays strong diffraction lines for antimony oxide Sb_6O_{13}, unreacted $CaCl_2$, and $CaSO_4$. This result indicates that the antimony volatilization proceeds according to the global reaction (1), and that Sb_6O_{13} is a non-volatile antimony oxide. The formation of this oxide prevents the complete volatilization of antimony.

Effect of Temperature on the Chloridizing of Stibnite

Figure 4 shows the results of the chloridizing of stibnite with calcium chloride obtained by using a stoichiometric mixture of $Sb_2S_3/CaCl_2$. In this figure, the data are shown as fraction of weight loss as a function of time and temperature. It can be observed that temperature has an important influence on the rate of weight loss of the sample.

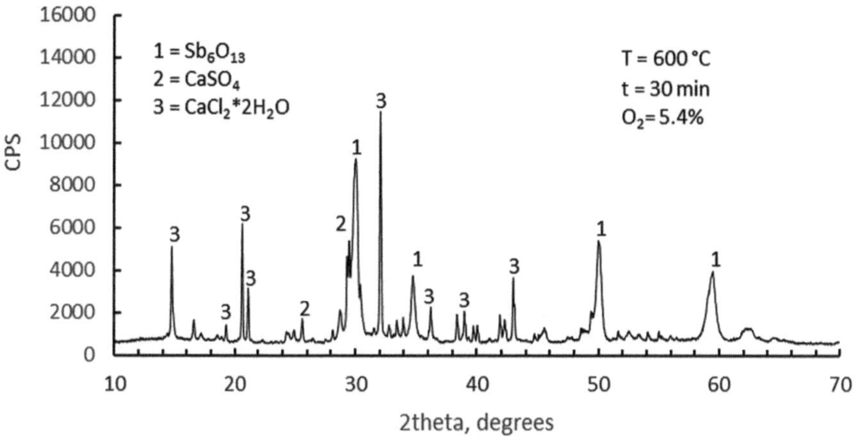

Fig. 3 XRD spectrum of a calcine produced using 50% excess of calcium chloride at the conditions shown in the figure

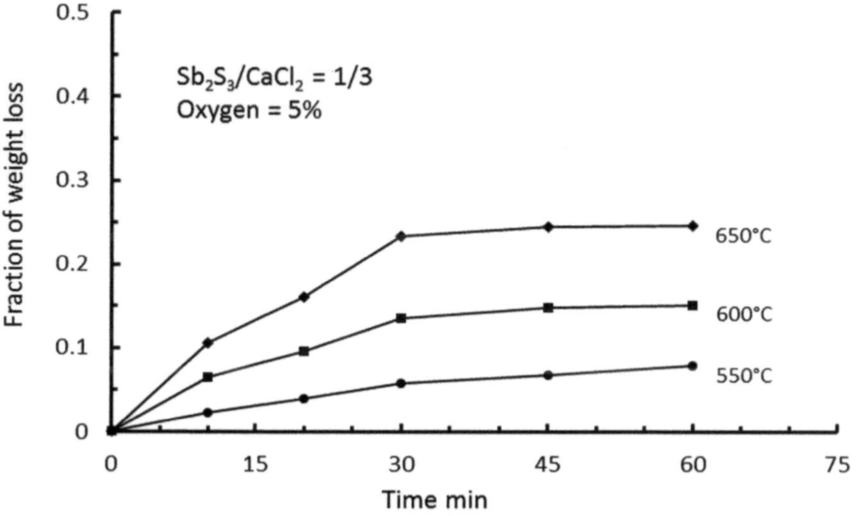

Fig. 4 Effect of temperature on the chloridizing of stibnite with calcium chloride

Based on the stoichiometry of reaction (1), the conversion of the chloridizing of stibnite was calculated from the weight loss data and the result is shown in Fig. 5 for a sample reacted in 1% oxygen.

We observe in this figure that the maximum conversion attained was about 64% at 650 °C, which is consistent with the formation of nonvolatile oxides preventing further volatilization of antimony.

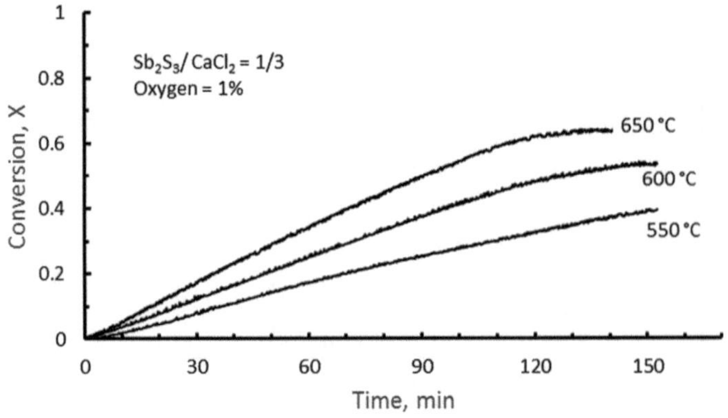

Fig. 5 Fractional conversion of stibnite as a function of temperature

Effect of the Partial Pressure of Oxygen

The influence of oxygen on the chloridizing of stibnite is shown in Fig. 6.

In this figure, we can observe that the rate of the antimony volatilization increases with increasing oxygen content in the gas phase. However, the maximum conversion diminishes with an increase in the oxygen content. Therefore, the highest conversion occurs when using 1% oxygen in the gas phase at 650 °C. The adverse effect of partial pressure of oxygen on the maximum conversion of stibnite is due to an increase in the rate of formation of nonvolatile antimony oxides.

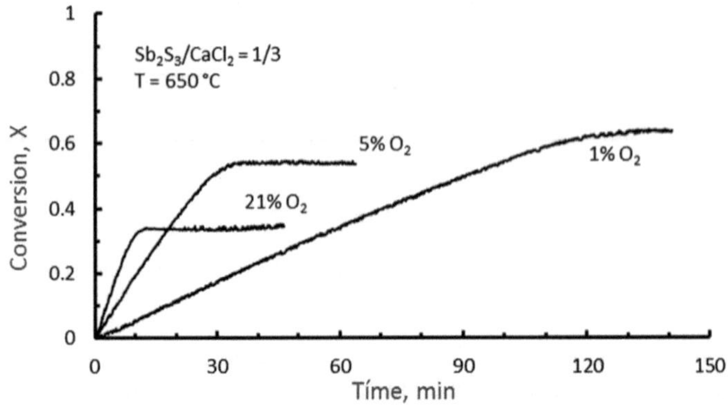

Fig. 6 Effect of the oxygen partial pressure of oxygen on the chloridizing of stibnite

Chloridizing of the MH Complex Copper Concentrate

Figure 7 shows the weight loss data for the chloridizing of a mixture of MH concentrate-$CaCl_2$ with a mass ratio (MH concentrate)/(Calcium chloride) = 10.5. The chloridizing roasting was carried out in the temperature range of 500 to 650 °C in 5% oxygen atmosphere. We can observe in this figure that temperature has an important effect on the rate of weight loss of the sample. For example at 600 °C and 5% oxygen, the weight loss was 14% in 20 min of reaction, while at the same conditions at 500 °C, the weight loss was 11%. It can be seen in the figure that each weight loss curve passes through a maximum and after that point, the sample gains weight due to the formation of highly oxidized compounds of the various sulfides, including copper sulfides, present in the concentrate.

Considering that enargite is one of the main components of the concentrate (22.4%), a major fraction of the weight loss of the sample is due to the chloridizing/volatilization of arsenic according to the global reaction (2):

$$Cu_3AsS_4 + 1.5CaCl_2 + 4O_2 = AsCl_3(g) + 1.5Cu_2S + 1.5CaSO_4 + SO_2(g) \quad (2)$$

In order to substantiate this reaction, two calcines produced at 600 °C in 5% oxygen and 2, and 10 min of reaction (see Fig. 7 2-min point marked with A) were analyzed by XRD spectroscopy. The results are shown in Figs. 8 and 9.

Figures 8 and 9 show the results of XRD analysis of a sample reacted for 2 and 10 min, respectively. It is interesting to note in Fig. 8 the absence of diffraction lines for enargite and pyrite. This implies that in two min of reaction time, enargite and pyrite reacted completely. The peaks identifying tennantite and the absence of peaks for enargite in Fig. 8 indicate that enargite was completely decomposed to tennantite. The absence of diffraction lines of pyrite indicates the formation of iron oxides.

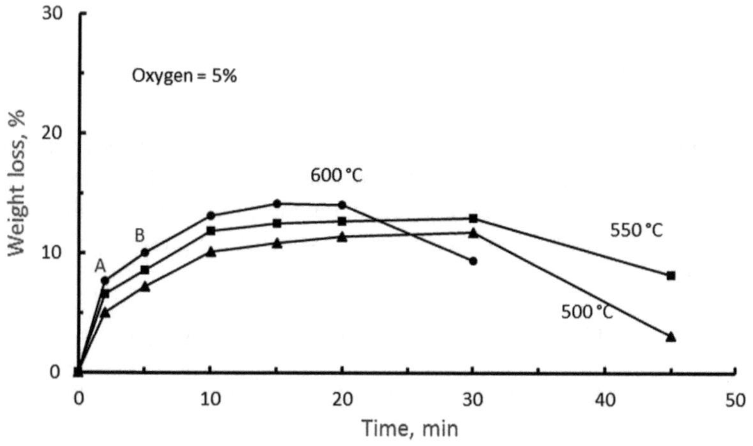

Fig. 7 Effect of temperature on the chloridizing of the complex copper concentrate

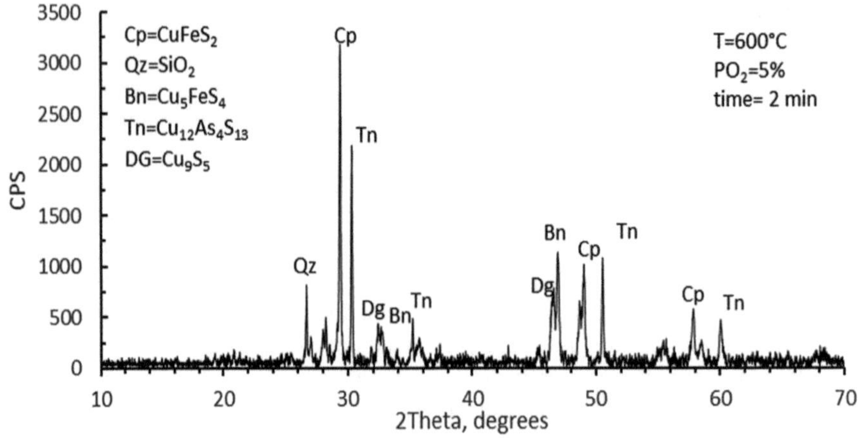

Fig. 8 XRD spectrum of a partially reacted sample at 600 °C for two min

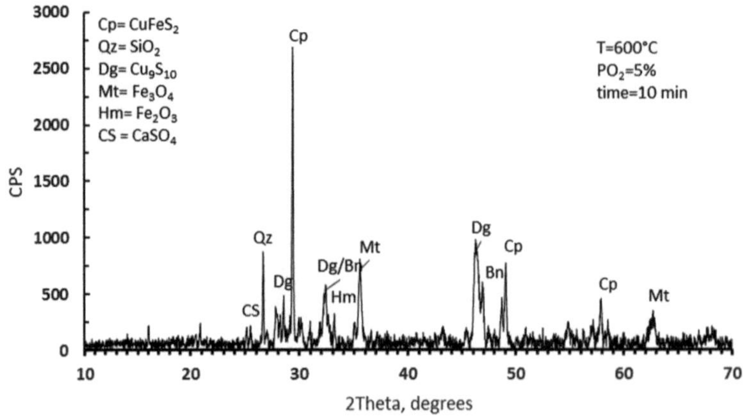

Fig. 9 XRD spectrum of a calcine obtained at 10 min at 600 °C and 5% oxygen in the gas phase

At 10 min (see Fig. 9), diffraction lines for arsenic compounds are absent and the lines for digenite and magnetite become significant and the formation of calcium sulfate is evident. Therefore, based on these XRD results the following reactions can be written for the behavior of arsenic in the presence of oxygen and $CaCl_2$:

$$4Cu_3AsS_4 + 3O_2(g) = Cu_{12}As_4S_{13} + 3SO_2(g) \quad (3)$$

$$3Cu_{12}As_4S_{13} + 18CaCl_2 + 37O_2(g) = 12AsCl_3(g) + 4Cu_9S_5 + 18CaSO_4 + SO_2(g) \quad (4)$$

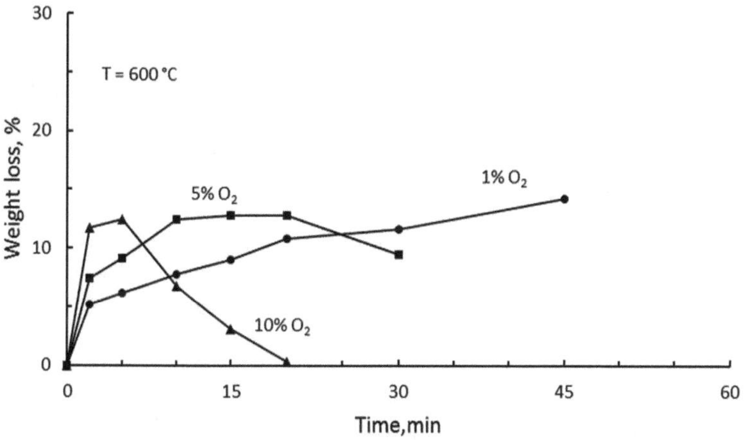

Fig. 10 Effect of the oxygen partial pressure in the gas phase on the chloridizing of the copper concentrate

These results suggest that arsenic is decomposed first to tennantite and subsequently tennantite is chloridized to arsenic trichloride plus the formation of digenite and calcium sulfate.

Effect of Oxygen Partial Pressure on the Chloridizing of MH Concentrate

The effect of partial pressure of oxygen on the chloridizing of the concentrate is shown in Fig. 10. We observe in this figure that oxygen concentration in the gas phase does affect not only the rate of weight loss due to volatilization of arsenic and other volatile species, but also the rate of formation of non-volatile compounds. It can also be observed that the weight loss curve shows a maximum, which depends on the oxygen concentration. For example at 10% oxygen concentration, the maximum occurs in less than 5 min, while for 5% oxygen in the gas phase, the maximum occurs at 20 min of reaction.

Conclusions

From the results of this study, the following can be concluded.

- Antimony volatilization from Sb_2S_3–$CaCl_2$ mixtures in the presence of oxygen at roasting temperatures (500–650 °C) proceeds according to the following reaction:

$$Sb_2S_3 + 3CaCl_2 + 6O_2(g) = 2SbCl_3(g) + 3CaSO_4$$

- The effect of calcium chloride concentration, molar ratio $Sb_2S_3/CaCl_2$ mixture is negligible on the rate of antimony volatilization.
- About 64% of antimony can be volatilized at 600 °C in 5% oxygen by roasting stibnite in the presence of calcium chloride.
- There is an adverse effect of partial pressure of oxygen on the stibnite chloridizing reaction due to the formation of nonvolatile antimony oxide Sb_6O_{13}.
- In the chloridizing of the complex copper concentrate, temperature has an important effect on the rate of weight loss of the sample. At 600 °C and 5% Oxygen, the weight loss was 14% in 20 min, while at the same conditions at 500 °C the weight loss was 11%.
- Based on the XRD results, the arsenic volatilization from the concentrate–calcium chloride mixture can be represented by:

$$4Cu_3AsS_4 + 3O_2(g) = Cu_{12}As_4S_{13} + 3SO_2(g)$$
$$3Cu_{12}As_4S_{13} + 18CaCl_2 + 37O_2(g) = 12AsCl_3(g) + 4Cu_9S_5 + 18CaSO_4 + SO_2(g)$$

- Oxygen concentration in the gas phase affects not only the rate of weight loss due to volatilization of arsenic and other volatile species, but also the rate of formation of oxidized non-volatile compounds.

Acknowledgements The authors acknowledge the National Fund for Scientific and Technological Development (FONDECYT) of Chile for the financial support of this study through the Project # 1150339.

References

1. Baxter K, Scriba H, Vega I (2010) Treatment of high-arsenic copper-gold concentrate—an options review. In: Proceedings of copper 2010, Germany, vol 5, pp 1783–1802
2. Padilla R, Ramirez G, Ruiz MC (2010) High temperature volatilization mechanism of stibnite in nitrogen-oxygen atmosphere. Metall Mater Trans B 41(6):1284–1292
3. Padilla R, Villa R, Ruiz MC, Reddy R (2011) Kinetics of the oxidation of bismuthinite in oxygen-nitrogen atmospheres. Metall Mater Trans B 42(5):1017–1022
4. Padilla R, Aracena A, Ruiz MC (2012) Reaction mechanism and kinetics of enargite oxidation at roasting temperatures. Metall Mater Trans B 43(5):1119–1126
5. Roca A et al (2003) A leaching process for removing arsenic from enargite-bearing copper concentrates. In: Riveros et al (eds) Proceedings of copper 2003, pp 631–644
6. Ruiz MC, Bello R, Padilla R (2013) Removal of arsenic from enargite rich copper concentrates. In: L Zhan et al (eds) Materials processing fundamentals. Wiley, Hoboken, pp 217–223
7. Holmstrom A (1988) Possibilities for the removal of antimony, arsenic and bismuth from concentrates rich in copper, silver and gold. Scand J Metall 17:248–258

Research on Sulfur Conversion Behavior in Oxygen Pressure Acid Leaching Process of High Indium Sphalerite

Yan Liu, Yang-yang Fan, Jun-fu Qi, Lei Tian and Ting-an Zhang

Abstract As an important treatment method of processing the sphalerite, the technology of pressure acid leaching has many benefits, such as process flow short, no waste pollution, low production costs, high leaching rate of Zn, and S output in elemental form, which can be sold as a product. This paper takes the high indium sphalerite provided by Kunming Metallurgical Research Institute as the research object. A metallurgical behavior of sulfur conversion process is studied under high temperature and high pressure. The results indicate that the conversion rate of sulfur increases with the increasing liquid-solid ratio, leaching temperature, initial acid concentration and partial oxygen pressure in a certain range. With increasing leaching time, the conversion rate of sulfur increases first and then decreases. Considering all the aspects, the optimum parameters in the condition of conventional electric heating and microwave heating of the process are as follows: liquid-solid ratio 8:1, leaching temperature 423 K, initial acid concentration 120 g/L, partial oxygen pressure 1.0 Mpa, leaching time 90 min. The conversion rate of sulfur reaches 72.00% under the condition of conventional electric heating, 65.74% under the condition of microwave heating.

Y. Liu (✉) · Y. Fan · J. Qi · L. Tian · T. Zhang
Key Laboratory of Ecological Metallurgy of Multi-metal Intergrown Ores of Ministry of Education, Special Metallurgy and Process Engineering Institute, Northeastern University, 110819 Shenyang, Liaoning, China
e-mail: shanqibao2000@163.com

Y. Fan
e-mail: 1047564874@qq.com

J. Qi
e-mail: 1432254582@qq.com

L. Tian
e-mail: 317660913@qq.com

T. Zhang
e-mail: zta2000@163.net

G. Lambotte et al. (eds.), *Materials Processing Fundamentals 2018*,
The Minerals, Metals & Materials Series,
https://doi.org/10.1007/978-3-319-72131-6_18

Keywords Conversion rate of sulfur · High indium sphalerite · Conventional electric heating · Microwave heating

Introduction

In the process of oxygen pressure leaching of sulphide, the problems of sulfur conversion and acid balance directly affect the metal balance, volume balance and slag-metal equilibrium of zinc smelting system [1–3]. Therefore, this is great practical significance to solve the problems of sulfur conversion and acid balance. There are two sources of sulfuric acid required for industrial production. One is the free acid from the liquid waste of zinc electrolysis, and the other is the sulfuric acid formed during the reaction process and the residual acid left after the reaction. The consumption of sulfuric acid can be divided into two categories. On the one hand, zinc concentrate contains zinc, iron, copper, cadmium, nickel, cobalt and so on. These elements can react with sulfuric acid to produce soluble sulfate. Besides, sulphuric acid is consumed in reaction, but the concentration of sulfate radical ions doesn't change. On the other hand, zinc concentrate also includes plumbum, calcium, barium and so on. However, these elements can react with sulfuric acid to form insoluble sulfate. This will result in the consumption of sulfuric acid, reducing of the concentration of sulfate radical ions and the loss of sulfur. So the essence of sulfur conversion and acid balance is the self-balancing for sulfur in the system [4–6].

Because of various valence state, there are particular conversion rules for sulfur under high pressure. It usually exists in the form of SO_4^{2-}, S, HS^-, S^{2-}, two of which, SO_4^{2-} and S are important raw materials and products of leaching process respectively [7–9]. Control the conversion rate of elemental sulfur and the oxidation rate of elemental sulfur by changing the leaching temperature, partial oxygen pressure, initial acid concentration and other leaching process conditions. The direction of sulfur is controlled in zinc concentrate. This can achieve the self-balancing for sulfur and unnecessary extra sulfuric acid under high pressure [10, 11].

The solution to sulfur conversion and acid balance requires to study the behavior of sulfur in oxygen pressure acid leaching process for zinc concentrate [12].

Experimental

Materials

This paper takes the high indium sphalerite provided by Kunming Metallurgical Research Institute as the research object. X-ray diffraction (XRD) pattern and SEM photograph of the high indium sphalerite is given in Fig. 1. The chemical composition is shown in Table 1.

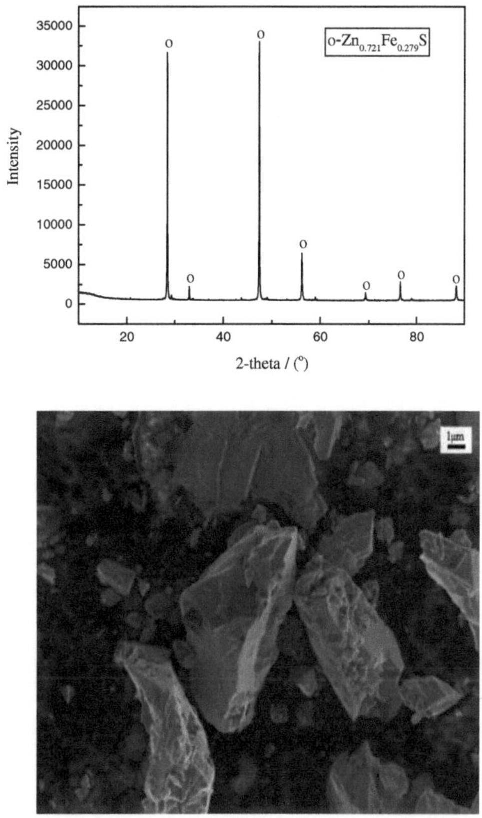

Fig. 1 XRD analysis of high indium sphalerite sample and photo of SEM

Table 1 Chemical composition of high indium sphalerite sample (%)

Component	Zn	S	Fe	Si	Mg	Pb	Ca
wt%	50.60	36.2	12.70	0.82	0.39	0.068	0.286
Cu	Al	In	Mn	As	Sn	K	Bi
0.882	0.217	0.094	0.14	0.215	0.075	0.002	<0.01

Leaching Experiment

The experimental equipment is HY-GF1501 microwave and electric hybrid heating autoclave (Hunan Huae Microwave Technology Co., Ltd. Changsha, China), which has three heating modes, such as pure electric heating, pure microwave heating and microwave and electric hybrid heating. It is shown in Fig. 2.

A certain amount of high indium sphalerite and calcium lignosulphonate was added at one time to the agitated 600 ml sulfuric acid solution of the required concentration when temperature reached the pre-set, and mixture was stirred at

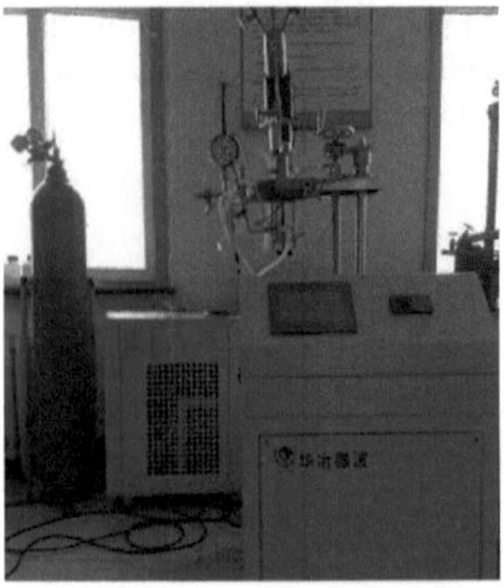

Fig. 2 HY-GF1501 microwave and electric hybrid heating autoclave

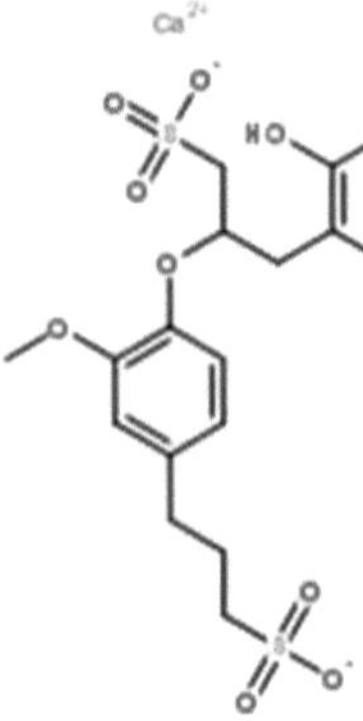

Fig. 3 Molecular structure of calcium lignosulfonate

500 rpm. Additive amount of calcium lignosulfonate was one percent weight of high indium sphalerite. The molecular structure of calcium lignosulfonate is given in Fig. 3. At set temperature, oxygen was admitted and partial oxygen pressure was adjusted to the desired value. After a certain amount of time, the leaching solution was withdrawn and the concentration of each element was analyzed by an inductive coupled plasma emission spectrometer.

The conversion rate of sulfur is calculated by Eq. (1).

$$\eta_S = \frac{(m_{Zn}/M_{Zn} + m_{Fe}/M_{Fe}) \times M_S}{m_S} \times 100\% \quad (1)$$

where η_S is the conversion rate of sulfur; m_{Zn} the content of zinc in the filtrate, g; M_{Zn} is the molar mass of zinc, g/mol; m_{Fe} the content of iron in the filtrate, g; M_{Fe} is the molar mass of iron, g/mol; m_S the total content of sulfur in the high indium sphalerite, g; M_S is the molar mass of sulfur, g/mol.

All the experiments were based on high indium sphalerite of particle size 48–75 μm. Experiments were performed with liquid-solid ratio of 8:1, except those relating to the effect of liquid-solid ratio on the conversion rate of sulfur. Liquid–solid ratio is the ratio of absolute solution volume to sphalerite quality. The leaching temperature, partial oxygen pressure, initial acid concentration and leaching time were fixed at 423 K, 1.0 MPa, 120 g L^{-1}, and 90 min, respectively.

Results and Discussion

Effect of Liquid-Solid Ratio on the Conversion Rate of Sulfur

The effect of liquid-solid ratio on the conversion rate of sulfur was investigated using different liquid-solid ratio (8:1, 6:1, 4:1, 2:1). The results of the leaching are shown in Fig. 4.

Figure 4 shows the conversion rate of sulfur enhances with increasing liquid-solid ratio. The conversion rate of sulfur is similar with conventional heating or microwave heating. For example, the conversion rate of sulfur reached 72.00% under conventional heating at liquid-solid ratio of 8:1; and the conversion rate of sulfur just 65.74% under microwave heating at liquid-solid ratio of 8:1. Compared with actual industrial conditions, it is considered that liquid-solid ratio of 8:1 is the priority.

Effect of Leaching Temperature on the Conversion Rate of Sulfur

The effect of leaching temperature on the conversion rate of sulfur was investigated using different leaching temperature (363, 383, 403, 423 K). The results of the leaching are shown in Fig. 5.

Fig. 4 Effect of liquid-solid ratio on conversion rate of S

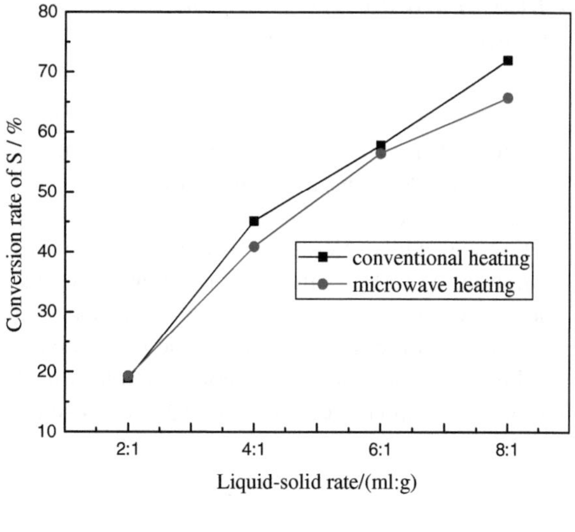

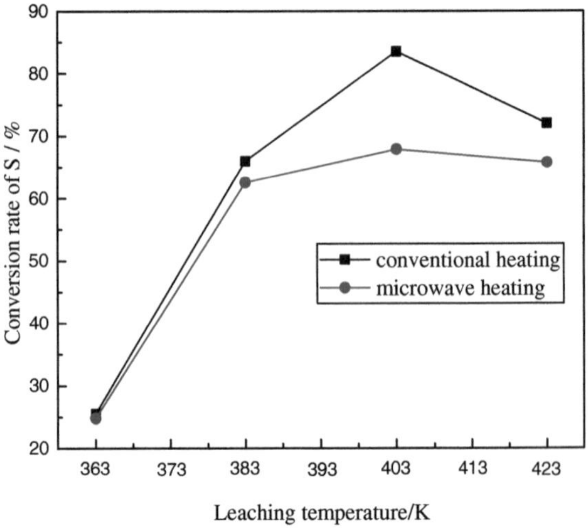

Fig. 5 Effect of leaching temperature on conversion rate of S

The result shows that leaching temperature has a remarkable effect on the conversion rate of sulfur. At 403 K, 83.51% of sulfur under conventional heating and 67.88% of sulfur under microwave heating were converted. However, these values decreased to 72.00 and 65.74% at 423 K, respectively. This is probably because sulfur reacts with oxygen and water to form sulfuric acid with increasing temperature. In general, increasing of leaching temperature, increase conversion rate of sulfur under conventional heating. This is more significant than that under microwave heating. In consideration of leaching of principal element of Zn, leaching temperature of 423 K is the priority.

Effect of Initial Acid Concentration on the Conversion Rate of Sulfur

The effect of initial acid concentration on the conversion rate of sulfur was investigated using different initial acid concentration (60, 90, 120, 150 g/L). The results of the leaching are shown in Fig. 6.

As we can see from Fig. 6, the conversion rate of sulfur both under conventional heating and under microwave heating increase with increasing of initial acid concentration. At 150 g/L, the conversion rate of sulfur under conventional heating and under microwave heating reached 77.45 and 76.01%, respectively. Due to the excessive sulfuric acid, it leads to the concentration of Fe^{3+} increasing sharply in leaching solution, and makes more difficulties in the next purification process. So the optimum process condition of 120 g/L is selected.

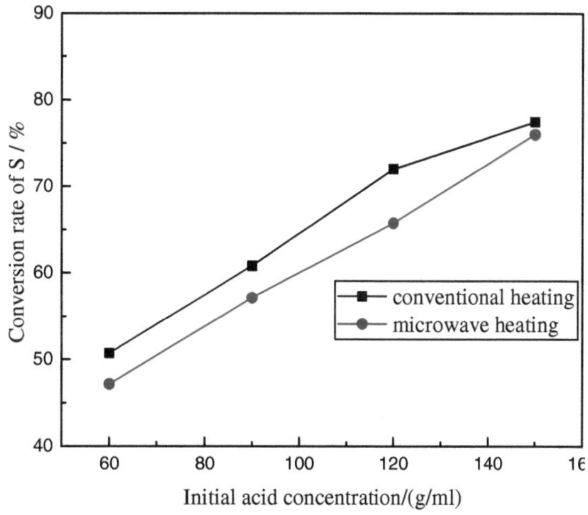

Fig. 6 Effect of initial acid concentration on conversion rate of S

Effect of Leaching Time on the Conversion Rate of Sulfur

The effect of leaching time on the conversion rate of sulfur was investigated using different leaching time (30, 60, 90, 120 min). The results of the leaching are shown in Fig. 7.

As shown from Fig. 7, with increasing of leaching time, the conversion rate of sulfur increases first and then decreases. At 60 min, 70.99% of sulfur under conventional heating and 66.42% of sulfur under microwave heating were converted. However, these values increased to 72.00 and 65.74% at 90 min, respectively. When time longer than 90 min are considered, sulfur reacts with oxygen and

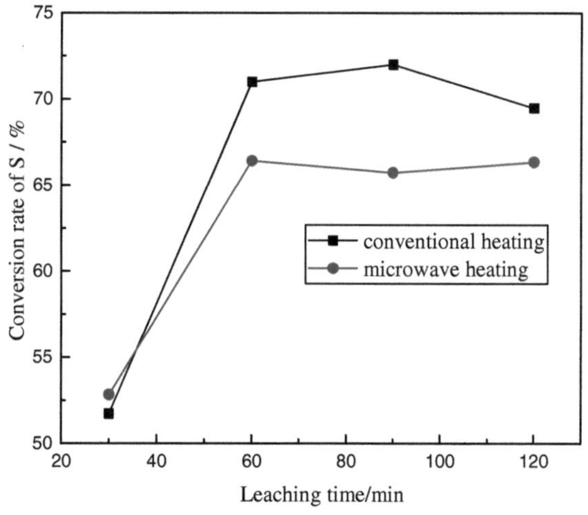

Fig. 7 Effect of leaching time on conversion rate of S

water to form sulfuric acid, causing a decline in the conversion rate of sulfur. In view of leaching of principal element of Zn, leaching time of 90 min is the priority.

Effect of Partial Oxygen Pressure on the Conversion Rate of Sulfur

The effect of partial oxygen pressure on the conversion rate of sulfur was investigated using different partial oxygen pressure (0.4, 0.6, 0.8, 1.0 Mpa). The results of the leaching are shown in Fig. 8.

Figure 8 shows that increasing of partial oxygen pressure, increases the conversion rate of sulfur gradually. At 0.6 Mpa, the conversion rate of sulfur under conventional heating and under microwave heating reached 73.75 and 69.37%, respectively. But the conversion rate of sulfur decreased after the partial oxygen pressure reaching to 0.6 Mpa. This is possibly due to sulfur reacts with oxygen and water to form sulfuric acid with increasing partial oxygen pressure. Besides, the result shows that conventional heating is better than microwave heating. In consideration of leaching of principal element of Zn, partial oxygen pressure of 1.0 Mpa is the priority.

SEM Micro-Morphology Analysis of Leaching Residue

The slag samples in the optimum process conditions of sulfur conversion were analyzed by SEM, one under conventional heating and the other under microwave heating. The results are as shown in Figs. 9 and 10, respectively.

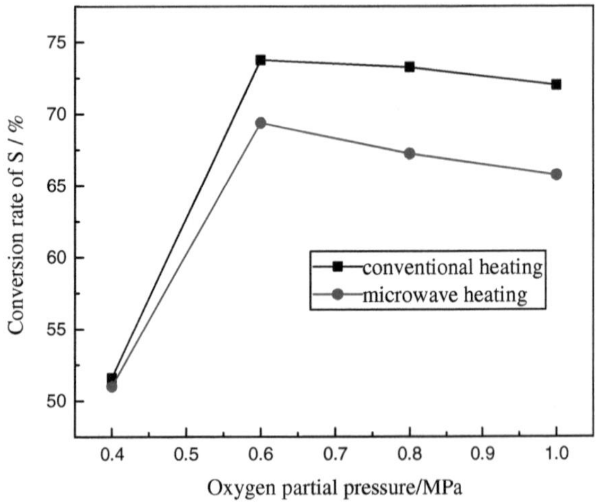

Fig. 8 Effect of partial oxygen pressure on conversion rate of S

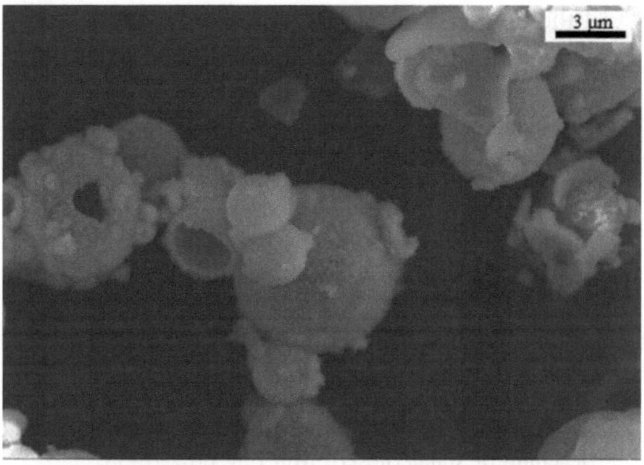

Fig. 9 Electronic microscope photograph of slag sample under microwave heating

Fig. 10 Electronic microscope photograph of slag sample under conventional heating

Figures 9 and 10 show the slag sample under conventional heating has a lot of damaged sulfur layers, but there are still some undamaged sulfur layers under microwave heating. The effect of microwave radiation on the conversion rate of sulfur is not significant, and its detailed mechanism need to be further analyzed.

Conclusion

The optimum parameters in the condition of conventional electric heating of the process are as follows: liquid–solid ratio 8:1, leaching temperature 423 K, initial acid concentration 120 g/L, partial oxygen pressure 1.0 Mpa, leaching time 90 min. And the conversion rate of sulfur reaches 72.00%.

The optimum parameters in the condition of microwave heating of the process are as follows: liquid-solid ratio 8:1, leaching temperature 423 K, initial acid concentration 120 g/L, partial oxygen pressure 1.0 Mpa, leaching time 90 min.

And the conversion rate of sulfur only reaches 65.74%. Furthermore, the detailed mechanism of microwave radiation need to be further analyzed.

Acknowledgements This research was supported by the National Natural Science Foundation of China (No.: U1402271).

References

1. Tian L, Liu Y, Lv GZ et al (2016) Research on sulphur conversion and acid balance from marmatite in pressure acid leaching[J]. Can Metall Q 55(4):438–447
2. Li ZH, Wang JK (2008) Study of sulphur recycling from the leaching slag of oxygen pressure leaching of sphalerites[J]. Min Eng 6(6):31–38
3. Long XY, Long K (2008) Reclaiming technology of sulfur from zinc sulfide concentrate leaching residue[J]. Inorg Chem Ind 40(10):53–61
4. Jin XZ, Yang ZP, Shi YH et al (2005) Experimental study of sulphur recycling from zinc blende concentrate in oxygen pressure acid leaching slag[J]. Min Resour Geol 19(1):97–104
5. Deng ML (2008) Formation mechanism of sulfur and sulfur recovery process in the process of pressure leaching of zinc sulfide concentrate[J]. Eng Des Res 2008(2):14–20
6. Deng RZ, Zhao TC (1992) Investigation on recovery of elemental sulfur from sulfur leaching residues[J]. J Central S Univ Technol (Nat Sci) 23(3): 287–294
7. Baldwin SA, Demopoulos GP, Papangelakis VG (1995) Mathematical modeling of the zinc pressure leach process[J]. Metall Mat Trans B 26(5):1035–1047
8. Perez IP, Dutrizac J (1991) The effect of the iron content of sphalerite on its rate of dissolution in ferric sulphate and ferric chloride media[J]. Hydrometallurgy 26(2):211–232
9. Piao SY, Tozawa K (1985) Effect of iron content in zinc sulfide concentrates on zinc extraction in oxygen pressure leaching with elemental sulfur[J]. J Min Metall Inst Jpn 101 (1174):795–800
10. Akcil A, Ciftci H (2002) A study of the selective leaching of complex sulphides from the Eastern Black Sea Region, Turkey[J]. Min Eng 15(6):457–459
11. Akcil A, Ciftci H (2003) Metals recovery from multimetal sulphide concentrates ($CuFeS_2$–PbS–ZnS): combination of thermal process and pressure leaching[J]. Int J Min Process 71(1–4):233–246
12. Sahu S, Sahu K, Pandey B (2006) Leaching of zinc sulfide concentrate from the ganesh-himal deposit of nepal[J]. Metall Mater Trans B 37(4):541–549

Part VI
Poster Session

Hybrid Modeling for Endpoint Carbon Content Prediction in EAF Steelmaking

Guang-sheng Wei, Rong Zhu, Lingzhi Yang and Tianping Tang

Abstract Considering the complicated and harsh conditions in the electric arc furnace (EAF) steelmaking process, the precise endpoint control technology is a crux that influences the product quality and production costs of the molten steel because precise endpoint control can control the endpoint carbon content and the endpoint oxidation. In this paper, a new hybrid prediction model was established to predict the endpoint carbon content in EAF steelmaking, which included the mechanism model based on the mass transfer process and the Extreme Learning Machine (ELM) optimized by the Evolving Membrane Algorithm (EMA). The mechanism model was calibrated with corrected parameters obtained from the ELM-EMA algorithm. As a result, the shortages that the mechanism model can't work precisely and that the single mathematical algorithm model lacks the analysis of the metallurgy process were overcome by the hybrid prediction model. Meanwhile, modifying ELM algorithm by EMA algorithm can improve the generalization performance of single-hidden-layer feed-forward neural networks. The experiments on a 50t EAF demonstrated that the proposed model had a good generalization performance and good prediction accuracy.

Keywords Hybrid model · Electric arc furnace · Extreme learning machine
Evolving membrane algorithm · Endpoint carbon content

G. Wei (✉) · R. Zhu · T. Tang
Metallurgical and Ecological Engineering School, University of Science and Technology Beijing, 100083 Beijing, China
e-mail: ustbwgsh@126.com

G. Wei · R. Zhu
Beijing Key Laboratory of Special Melting and Preparation of High-End Metal Materials, 100083 Beijing, China

L. Yang
School of Minerals Processing and Bioengineering, Central South University, 410083 Changsha, China

G. Lambotte et al. (eds.), *Materials Processing Fundamentals 2018*,
The Minerals, Metals & Materials Series,
https://doi.org/10.1007/978-3-319-72131-6_19

Introduction

As one of the main production methods in modern iron and steel industry, electric arc furnace (EAF) steelmaking offers the secondary metallurgic processes molten steel with required composition and temperature [1]. Nowadays, enhanced oxygen technology is adopted to reduce the energy consumption and promote the smelting rhythm in EAF steelmaking. However, the increasing amount of oxygen leads to a series of problems, such as the low metal yield, the sharp increase of inclusions, the oxidative damage of the conductive cross arms of EAF and so on. Therefore, precise prediction of the endpoint carbon content is essential for EAF steelmaking.

The physical and chemical interactions in the process form a complex system of high temperature, multi phase and high reaction speed. Considering the complicated and harsh conditions, some parameters can't be measured accurately in actual production. With the development of automation technology, the soft sensor technology is widely used for the online measurement of some key parameters in industrial processes [2]. Currently, the soft sensor technology for endpoint carbon content prediction of EAF steelmaking mainly includes mechanism models and intelligent algorithm models ("black box" models) [3]. Limited by detection techniques, the mechanism models show low prediction accuracy for adopting some empirical data. As to the "black box" models, variables relevant to the endpoint carbon content are simply inputted into the model and then the results can be obtained directly [4]. The shortage of these "black box" models is the limited prediction for lacking the analysis of the metallurgy process. Extreme Learning Machine (ELM) is a learning algorithm for single-hidden-layer feed-forward neural networks [5]. The traditional classic artificial neural network algorithms may face several issues like low speed, minima, over-fitting and so on [6]. The ELM can avoid these issues and obtains the global optimal solution with fast learning speed. Based on P systems, Evolving Membrane Algorithm (EMA) is an optimization technology to solve the numerical optimization problems and it exhibits a good ability of global search and convergence speed [7]. In this paper, EMA was employed to optimize the input weights and thresholds of ELM to promote the global optimization performance.

In this paper, the complicated physicochemical processes and the mass transfer process of carbon in EAF steelmaking were analyzed. Combining the traditional mechanism modeling method and the single intelligent modeling method, a new hybrid intelligent prediction model for endpoint carbon content in EAF steelmaking was built with ELM and EMA. Not only did the new prediction model overcome the difficulty of obtaining the parameters in the mechanism model but also it avoided the shortage of lacking the analysis of the EAF metallurgy process using the single intelligent algorithm model.

Mechanism Model for Endpoint Carbon Content Prediction

The EAF steelmaking process is a complicated physicochemical process with high temperature [8]. In this paper, the steelmaking process of a 50t EAF was studied and the mechanism model for endpoint carbon content prediction was built by analyzing the material balance and the relevant chemical reactions.

Input Model of Carbon

Taking the domestic price of electricity, scrap reserves, production costs and other factors into account, the hot metal charging technology is widely adopted in China [9].

According to the steelmaking process of the 50t EAF, the transmission of carbon has been studied and the source of carbon in the process has been found out, which is shown in Table 1 in details.

(1) Raw materials. The added materials, like molten iron, scrap and alloy materials, are rich in carbon and they are the main source of carbon in EAF steelmaking.
(2) Powdered carbon. During the smelting process, powdered carbon is injected into the furnace tank to form foamed slag which can improve the smelting efficiency.
(3) Electrode dissipation. The end and side parts of electrodes can be damaged and bring about the increase of carbon content in the molten steel.

Hence, the Input Model of Carbon in EAF Steelmaking is shown as the following expression.

$$M_{Cinput} = M_{molten} \cdot C_{molten\text{-}c} + M_{scrap} \cdot C_{scrap\text{-}c} + M_{coke} \cdot C_{coke\text{-}c} \cdot W_1 + M_{powder\text{-}c} \cdot C_{powder\text{-}c} \cdot \rho + M_{DRI} \cdot C_{DRI\text{-}c} + M_{electrode} \cdot C_{electrode\text{-}c} + M_X \quad (1)$$

where, M_{Cinput} is the total amount of carbon fusing into the molten steel. M_{molten}/M_{scrap}/M_{coke}/$M_{powder\text{-}c}$/M_{DRI}/$M_{electrode}$ is the mass of Molten iron/Scrap/Coke/

Table 1 The source of carbon in EAF steelmaking process

	Mass/kg	Percentage of carbon	Mass of carbon/kg
Molten iron	M_{molten}	$C_{molten\text{-}c}$	$M_{molten} \cdot C_{molten\text{-}c}$
Scrap	M_{scrap}	$C_{scrap\text{-}c}$	$M_{scrap} \cdot C_{scrap\text{-}c}$
Coke	M_{coke}	$C_{coke\text{-}c} \cdot W_1$	$M_{coke} \cdot C_{coke\text{-}c} \cdot W_1$
Powdered carbon	$M_{powder\text{-}c}$	$C_{powder\text{-}c} \cdot \rho$	$M_{powder\text{-}c} \cdot C_{powder\text{-}c} \cdot \rho$
DRI	M_{DRI}	$C_{DRI\text{-}c}$	$M_{DRI} \cdot C_{DRI\text{-}c}$
Electrode dissipation	$M_{electrode}$	$C_{electrode\text{-}c}$ ($\approx$99%)	$M_{electrode} \cdot C_{electrode\text{-}c}$
Alloy materials	FeCr, FeMn, SiMn, etc.		M_x

Powdered carbon/DRI/Electrode dissipation. $C_{molten\text{-}c}/C_{scrap\text{-}c}/C_{coke\text{-}c}/C_{powder\text{-}c}/C_{DRI\text{-}c}/C_{electrode\text{-}c}$ is the content of carbon in Molten iron/Scrap/Coke/Powdered carbon/DRI/Electrode dissipation. W_1 is the recycling utilization rate of carbon in coke. ρ is the percentage of powdered carbon fusing into molten steel. M_x is the approximate mass of carbon in alloy materials.

Model of Oxygen Utilization

On the basis of the material balance and the chemical equilibrium, the model of oxygen utilization in EAF steelmaking includes the input model of oxygen and the depletion model of oxygen [10]. During the smelting process, oxygen gas is injected into EAF by the oxygen lances and air continuously getting into the furnace tank may provide a certain amount of oxygen to participate in chemical reactions. In addition, at high temperatures, the desulphurization reaction (CaO + [S] = CaS + [O]) occurs and CaO in the lime stone will react with S to form oxygen in the molten pool. The remaining oxygen gas is captured by the exhaust, which is the off-gas flue of EAF and this part is not included in the input model of oxygen. Table 2 shows the source of oxygen in EAF steelmaking.

Hence, the input model of oxygen in EAF steelmaking is shown as the following expression.

$$M_{Oinput} = M_{oxy\text{-}gas} \cdot C_{oxy\text{-}gas} \cdot W_2 + M_{O\text{-}Air} + M_{i\text{-}ore} \cdot C_{Fe_2O_3} \cdot 48/160 + M_{lime} \cdot C_S \cdot 16/32 \tag{2}$$

where, M_{Oinput} is the total amount of oxygen inputted into the molten steel, $M_{oxy\text{-}gas}/M_{O\text{-}Air}/M_{i\text{-}ore}/M_{lime}$ is the mass of oxygen gas injected by oxygen lances/the mass of oxygen supplied by air/iron ore added/lime stone added, $C_{oxy\text{-}gas}$ is the percentage of oxygen in oxygen gas, $C_{Fe_2O_3}$ is the percentage of Fe_2O_3 in iron ore, C_S is the percentage of S in lime stone and W_2 is the utilization ratio of oxygen gas.

After a series of chemical reactions, oxygen element mainly exists in CO, CO_2 and unreacted O_2, and the rest exists in the liquid steel and the oxides of Fe, Si, Mn and P in the slag. Table 3 shows the concerned oxidation reactions and the consumption of oxygen.

Table 2 The source of oxygen in EAF steelmaking process

	Mass of Oxygen/kg*
Oxygen gas	$M_{oxy\text{-}gas} \cdot C_{oxy\text{-}gas} \cdot W_2$
Air	$M_{O\text{-}Air}$
Iron ore	$M_{i\text{-}ore} \cdot C_{Fe_2O_3}$ 48/160
Lime stone	$M_{lime} \cdot C_S \cdot$ 16/32

*Fraction ratio based on corresponding oxygen to oxide ratio

Table 3 The concerned oxidation reactions and the consumption of oxygen

Elements	Oxidation reactions	Mass of reacted elements/kg	Mass of consumed oxygen*
C	[C] → CO	M_{C1}	$M_{O1} = M_{C1} \cdot 16/12$
	[C] → CO_2	M_{C2}	$M_{O2} = M_{C2} \cdot 32/12$
Si	[Si] → (SiO_2)	M_{Si}	$M_{O\text{-}Si} = M_{Si} \cdot 32/28$
Mn	[Mn] → (MnO)	M_{Mn}	$M_{O\text{-}Mn} = M_{Mn} \cdot 16/55$
P	[P] → (P_2O_5)	M_P	$M_{O\text{-}P} = M_P \cdot 80/62$
Fe	[Fe] → (FeO)	M_{Fe1}	$M_{O\text{-}Fe1} = M_{Fe1} \cdot 16/56$
	[Fe] → (Fe_2O_3)	M_{Fe2}	$M_{O\text{-}Fe2} = M_{Fe2} \cdot 48/112$

*Fraction ratio based on corresponding oxygen to oxide mass ratio

During the smelting process, a portion of Fe reacts with oxygen to form FeO and Fe_2O_3 which are included in the smoke dust and the electric-furnace slag. The mass of Fe in FeO and Fe_2O_3, M_{Fe1} and M_{Fe2}, is respectively shown as the following expressions.

$$M_{Fe1} = \eta \cdot M_{molten} \cdot C_{molten-Fe} \cdot (1 - \omega) \cdot \alpha \tag{3}$$

$$M_{Fe2} = \eta \cdot M_{molten} \cdot C_{molten-Fe} \cdot \omega + \eta \cdot M_{molten} \cdot C_{molten-Fe} \cdot (1 - \omega) \cdot (1 - \alpha) \tag{4}$$

where, η is the burn out rate of Fe, ω is the proportion of Fe which is oxidized to become a part of the smoke dust and α is the proportion of Fe in the slag oxidized to FeO.

Therefore, the depletion model of oxygen is obtained by the above analysis.

$$M_{Ooutput} = M_{O1} + M_{O2} + M_{O-Si} + M_{O-Mn} + M_{O-P} + M_{O-Fe1} + M_{O-Fe2} \tag{5}$$

where, M_{Oinput} is the total amount of oxygen depleted, $M_{C1}/M_{C2}/M_{Si}/M_{Mn}/M_P/M_{Fe1}/M_{Fe2}$ is the mass of reacted element C/C/Si/Mn/P/Fe/Fe in oxide CO/CO_2/SiO_2/MnO/P_2O_5/FeO/Fe_2O_3 and $M_{O1}/M_{O2}/M_{O\text{-}Si}/M_{O\text{-}Mn}/M_{O\text{-}P}/M_{O\text{-}Fe1}/M_{O\text{-}Fe2}$ is the mass of consumed oxygen in oxide CO/CO_2/SiO_2/MnO/P_2O_5/FeO/Fe_2O_3.

M_{Si}, M_{Mn} and M_P are respectively obtained by the following expressions.

$$M_{Si} = M_{molten} \cdot C_{molten-Si} + M_{scrap} \cdot C_{scrap-Si} + M_X \cdot C_{X-Si} - M_{final} \cdot C_{final-Si} \tag{6}$$

$$M_{Mn} = M_{molten} \cdot C_{molten-Mn} + M_{scrap} \cdot C_{scrap-Mn} + M_X \cdot C_{X-Mn} - M_{final} \cdot C_{final-Mn} \tag{7}$$

$$M_P = M_{molten} \cdot C_{molten-P} + M_{scrap} \cdot C_{scrap-P} + M_X \cdot C_{X-P} - M_{final} \cdot C_{final-P} \tag{8}$$

where, $C_{molten\text{-}Si}/C_{scrap\text{-}Si}/C_{X\text{-}Si}$ is the content of Silicon in the molten iron/scrap/alloy materials, $C_{molten\text{-}Mn}/C_{scrap\text{-}Mn}/C_{X\text{-}Mn}$ is the content of Silicon in the molten iron/scrap/alloy materials, $C_{molten\text{-}P}/C_{scrap\text{-}P}/C_{X\text{-}P}$ is the content of Silicon in the molten iron/scrap/alloy materials, and $C_{final\text{-Si}}/C_{final\text{-Mn}}/C_{final\text{-P}}$ is the specified content of Si/Mn/P in the final molten steel.

The following expression is usually assumed in the mechanism model of EAF.

$$M_{Oinput} = M_{Ooutput} \tag{9}$$

Mechanism Model for Endpoint Carbon Content Prediction

According to the chemical equilibrium, the material balance and the supply of oxygen, the endpoint carbon content in the molten steel can be calculated. Based on the input model of oxygen and the depletion model of oxygen, the mass of oxygen reacting with carbon to form CO and CO_2 can be calculated with the following expression.

$$\begin{aligned} M_{O\text{-}C} &= M_{O1} + M_{O2} \\ &= M_{Ooutput} - (M_{O-\mathrm{Si}} + M_{O-\mathrm{Mn}} + M_{O-\mathrm{P}} + M_{O-\mathrm{Fe1}} + M_{O-\mathrm{Fe2}}) \\ &= M_{Oinput} - (M_{O-\mathrm{Si}} + M_{O-\mathrm{Mn}} + M_{O-\mathrm{P}} + M_{O-\mathrm{Fe1}} + M_{O-\mathrm{Fe2}}) \end{aligned} \tag{10}$$

The mass of carbon being oxidized to CO and CO_2 can be calculated with the following expression.

$$M_{C\text{-}Consume} = 3 \cdot M_{O\text{-}C}/(8 - 4\beta) \tag{11}$$

where, $\boldsymbol{\beta}$ is the proportion of carbon oxidized to CO among total carbon reacted and it can be figured out by flue gas monitoring.

Therefore, the mass of carbon in the final molten steel can be obtained by the above analysis.

$$M_{[C]} = M_{Cinput} - M_{C\text{-}Consume} \tag{12}$$

Most of the additions, like lime stone, powdered carbon and so on, will get into the slag finally. Hence, the mass of the final molten steel can be calculated by the following assumption.

$$M_{final} = M_{molten} + M_{scrap} + M_{DRI} \tag{13}$$

Eventually, the mechanism model for endpoint carbon content prediction in EAF steelmaking is shown as the following expression.

$$[C] = M_{[C]} \cdot 100/M_{final} \tag{14}$$

Based on the above analysis, it can be found that the main influence factors of these parameters includes the mass of C, Mn, Si, P and S in molten steel and scrap, the mass of the powdered carbon and the electrode dissipation, the mass of the oxygen gas supplied by lances and the power of the exhaust fan of the exhaust. Hence, these factors were selected as the input layer nodes of ELM.

Optimized ELM

Extreme Learning Machine (ELM)

A fast intelligent learning algorithm for single-hidden-layer feed-forward neural networks (SLFNs) called extreme learning machine (ELM) was proposed by Huang et al. [5]. Unlike other traditional neural networks, the input weights (linking the input layer to the hidden layer) and thresholds of ELM are randomly generated, and the output weights (linking the hidden layer to the output layer) are determined by using Moore-Penrose (MP) generalized inverse. Several issues of the traditional algorithms for SLFNs are avoided by ELM, such as local minima, improper learning rate, overfitting and so on. Compared to other traditional methods, ELM has a faster learning speed and a better generalization performance. The description of ELM is given as follows.

A typical structure of the ELM network is shown in Fig. 1. The input, hidden and output layers are connected by weights. As previous reported [5], the for a standard SLFNs with L hidden nodes, a standard SLFNs with L hidden nodes and activation function *g(x)* are mathematically modeled as:

$$\sum_{i=1}^{L} \boldsymbol{\beta}_i g_i(\boldsymbol{x}_j) = \sum_{i=1}^{L} \boldsymbol{\beta}_i g(\boldsymbol{w}_i \cdot \boldsymbol{x}_j + b_i) = \boldsymbol{o}_j, \quad j = 1, 2, \ldots, N \tag{15}$$

And the calculation formula is as follows:

$$\mathbf{H}\boldsymbol{\beta} = \mathbf{T} \tag{16}$$

Fig. 1 The structure of ELM

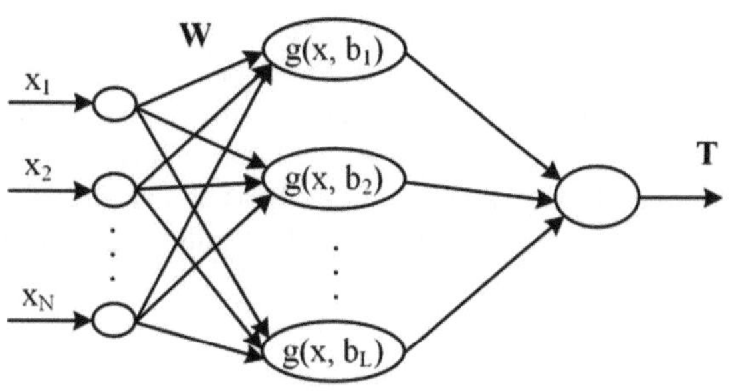

where

$$\mathbf{H} = \begin{bmatrix} g(\boldsymbol{w}_1 \cdot \boldsymbol{x}_1 + b_1) & \cdots & g(\boldsymbol{w}_L \cdot \boldsymbol{x}_1 + b_L) \\ \vdots & \ddots & \vdots \\ g(\boldsymbol{w}_1 \cdot \boldsymbol{x}_N + b_1) & \cdots & g(\boldsymbol{w}_L \cdot \boldsymbol{x}_1 + b_L) \end{bmatrix}_{N \times L} \tag{17}$$

$$\boldsymbol{\beta} = \begin{bmatrix} \boldsymbol{\beta}_1^T \\ \vdots \\ \boldsymbol{\beta}_L^T \end{bmatrix}_{L \times m} \tag{18}$$

and

$$\mathbf{T} = \begin{bmatrix} \boldsymbol{t}_1^T \\ \vdots \\ \boldsymbol{t}_N^T \end{bmatrix}_{N \times m} \tag{19}$$

where, **H** is called the hidden layer matrix of the SLFN.

The smallest norm least-squares solution of the above linear system is:

$$\hat{\boldsymbol{\beta}} = \mathbf{H}^{\dagger}\mathbf{T} \tag{20}$$

where $\mathbf{H}^{\dagger}$ is the Moore-Penrose generalized inverse of matrix **H**.

Thus, the output weight $\boldsymbol{\beta}$ can be calculated by the formula (20).

Evolving Membrane Algorithm (EMA)

Inspired by the structure and the function of the biological cells, Păun has proposed a novel distributed and parallel computing model, named as Membrane Computing [11]. Based on the membrane computing theory, a new optimized algorithm called evolving membrane algorithm (EMA) was proposed by Liu et al. [7]. In EMA, the reaction rules simulate the irregular Brownian motion to find the approximate solutions of the global numeric optimization problem. The global searching ability and the convergence performance of EMA are excellent. As Liu [7] reported, the EMA includes the following key steps: (1) Parameters initialization, (2) Creation of skin membrane and evolution rules, (3) Evaluation and classification of symbol-objects, (4) Evolution of the elementary membrane, (5) Global optimal solution of the optimization problems and acceleration of the convergence speed of EMA, (6) Cycle calculation and condition judgment and (7) Achievement of the globally optimal solution.

Optimized ELM Model

In the traditional ELM algorithm, the input weights and thresholds of ELM are both randomly generated. However, the global optimization and the calculation speed will be deteriorated for these parameters being adjusted repeatedly. Scholars have proposed many improvements to promote the generalization performance of ELM, such as OP-ELM and EM-ELM [12, 13]. In this paper, EMA is employed to produce the input weights and thresholds of ELM and then, the output weights of ELM can be obtained by Moore-Penrose. The new computational model is called optimized ELM model and Fig. 2 shows the structure of optimized ELM model [7].

The root mean square error (RMSE) between the output value of ELM and the real value is calculated as the fitness of symbol-objects in EMA. According to the RMSE value, EMA updates the information of the symbol-objects so that the parameters of ELM can be adaptively adjusted. The output of ELM is optimized continuously.

Hybrid Intelligent Prediction Model for Endpoint Carbon Content

Structure of Hybrid Intelligent Prediction Model

The mechanism model aforementioned shows the change of carbon content in the molten steel during the EAF smelting process. However, some key parameters, such as $M_{O\text{-}Air}$, η, α and ω, can't be obtained accurately in actual production just because of the complicated and harsh conditions. Hence, the prediction result of the

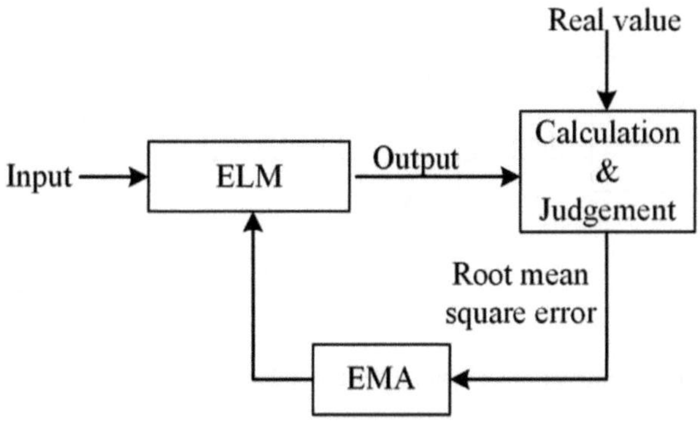

Fig. 2 The structure of optimized ELM model

endpoint carbon content of EAF steelmaking is inaccurate and as a result, the mechanism model can't be adopted to guide production for its low prediction accuracy. As to single mathematical algorithm models, these algorithms are employed to form "black box" without considering the metallurgical reaction mechanisms. Variables relevant to the endpoint carbon content are simply inputted into the model and then the result can be obtained directly [4]. The shortage of this method is the limited prediction for its only depending on the data and lacking the analysis of EAF metallurgy process.

Considering these shortcomings, a hybrid intelligent modeling method is proposed in this paper, which combining the intelligent algorithms and the mechanism model. Figure 3 shows the structure of the hybrid intelligent prediction model for endpoint carbon content in EAF steelmaking. The mechanism model and the optimized ELM are combined to predict the endpoint carbon content. Firstly, the ELM algorithm is employed to predict $M_{O\text{-}Air}$, η, α and ω of the mechanism model, and then these parameters obtained are inputted into the mechanism model to figure out the predicted value of endpoint carbon content. After that, according to the RMSE between the predicted carbon content and the real carbon content, the input weights and thresholds of ELM are modified by EMA. Therefore, further adjustment of relevant parameters in the mechanism model is achieved and a more accurate model is set up to predict the endpoint carbon content eventually. Not only does the hybrid intelligent model solve the problem that key parameters can't be obtained accurately in the mechanism model but also it overcomes the shortage that the single intelligent model has a poor generalization ability for lacking the analysis of the EAF metallurgy process. In this paper, the ELM algorithm is employed as the intelligent module of the hybrid intelligent model for its simple structure, good generalization ability and fast training speed. Meanwhile, the EMA algorithm is used for modifying relevant parameters of the ELM to promote the global optimization of the prediction model.

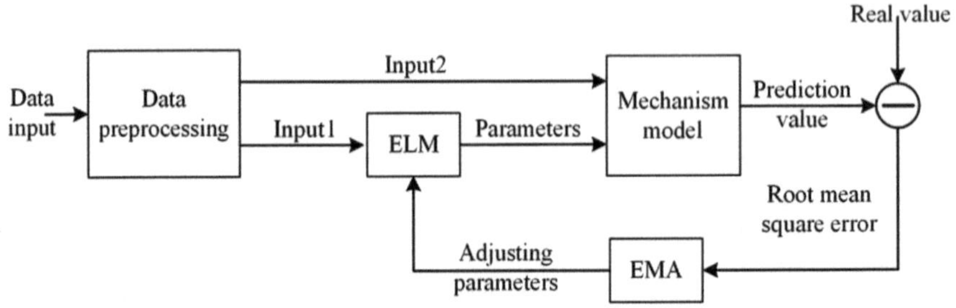

Fig. 3 The hybrid intelligent prediction model for endpoint carbon content in EAF steelmaking

Relevant Parameters Predicted by ELM in the Mechanism Model

In the mechanism modeling process, oxygen consumed during the EAF steel-making process is mainly supplied by oxygen gas, air and lime. Meanwhile, the amount of oxygen element from oxygen air and lime can be figured out by the amount of oxygen gas and lime supplied. Nevertheless, considering the poor tightness of the EAF, the ability of the exhaust for collecting smoke, air being ionized by electric arc and so on, it's difficult to measure the amount of oxygen from air in the process by effective detection methods. This part oxygen plays an important role in the endpoint carbon content prediction and it is a key link of the prediction model to obtain the accurate value of $M_{O\text{-}Air}$. Similarly, during the metallurgy process, Fe is burnt out to form FeO and Fe_2O_3. The relevant parameters (η, α and ω) is closely linked with the total amount of oxygen supplied, the content of C, Mn, Si, P and S, etc. However, it's hard to figure out them accurately. In general, these parameters are estimated by experience in traditional methods that will lead to low prediction accuracy. Therefore, the intelligent algorithm is adopted to predict $M_{O\text{-}Air}$, η, α and ω to enhance the accuracy of the model.

Experiments

In the hybrid intelligent model, the ELM algorithm is employed to predict $M_{O\text{-}Air}$, η, α and ω of the mechanism model. According to the above analysis, the main influence factors of these parameters includes the mass of C, Mn, Si, P and S in molten steel and scrap, the mass of the powdered carbon and the electrode dissipation, the mass of the oxygen gas supplied by lances and the power of the exhaust fan of the exhaust. Hence, the input layer of ELM has 9 nodes, the hidden layer has 20 nodes and the output layer has 4 nodes ($M_{O\text{-}Air}$, η, α and ω). Meanwhile, the EMA is used to optimize the input weights and thresholds of ELM with the RMSE of the training data. 550 sets of production data from the 50t EAF in January 2015 are adopted. They are divided into two groups. 500 sets of them are selected randomly to train the hybrid intelligent model, and the others are used to test. Figure 4 shows the endpoint carbon content predicted by the single ELM "black box" model and the hybrid model with ELM. It can be observed that the performance of the single ELM "black box" model is worse than the hybrid ELM model. Figure 5 shows the endpoint carbon content predicted by the hybrid model with ELM and the hybrid model with EMA-ELM. The prediction errors of the three methods are presented in Fig. 6. The results of experiments demonstrate that the hybrid model with EMA-ELM is better than the other two models. The comparison of performance of the relevant three models can be obtained by calculating the RMSE and the prediction accuracy as Table 4 shows. The prediction accuracy is calculated as follows:

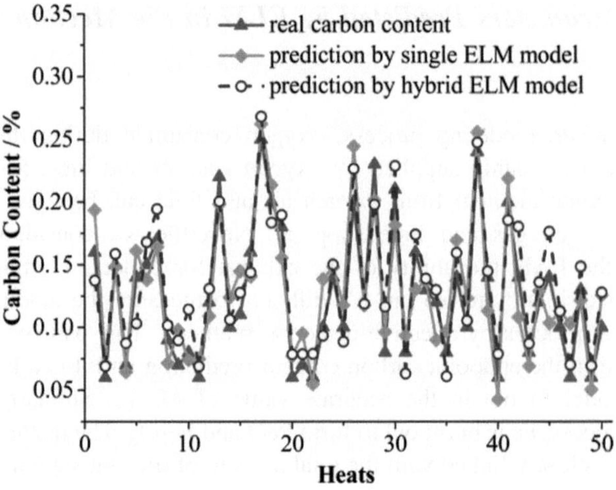

Fig. 4 The comparison of endpoint carbon content predicted by the single ELM "black box" model and the hybrid model with ELM

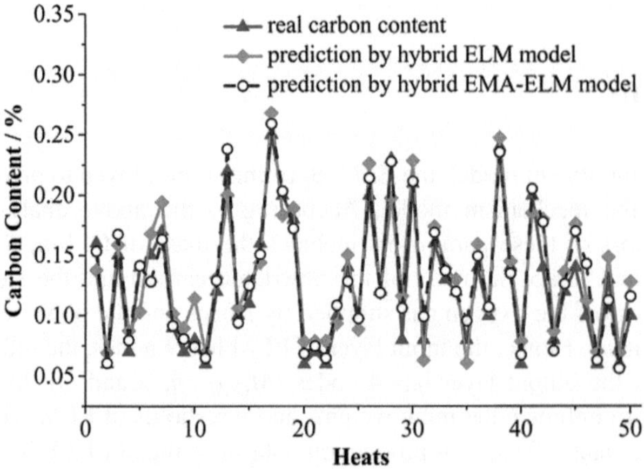

Fig. 5 The comparison of endpoint carbon content predicted by the hybrid model with ELM and the hybrid model with EMA-ELM

$$accuracy = \frac{N_C}{N_W} \tag{21}$$

where, N_C is the number of endpoint carbon content with error <0.02, N_W is the number of whole testing data.

According to the comparison above, it can be concluded that the hybrid intelligent model with EMA-ELM has a good generalization performance and good prediction accuracy. The prediction accuracy of the Hybrid intelligent model is up

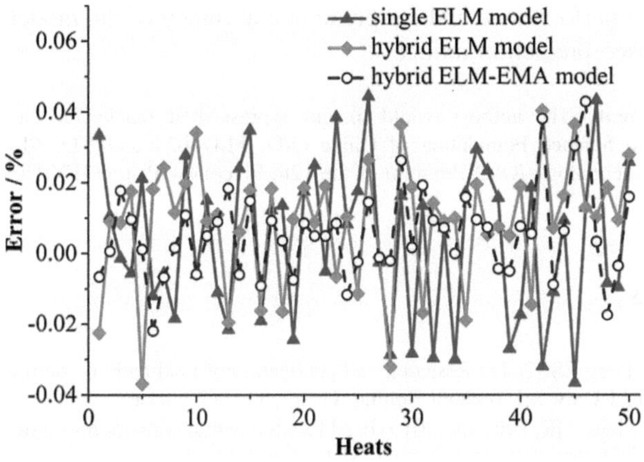

Fig. 6 The comparison of the errors of the single ELM "black box" model, the hybrid model with single ELM and the hybrid model with EMA-ELM

Table 4 Comparison of prediction by models proposed

	Single ELM "black box" model	Hybrid model with single ELM	Hybrid model with EMA-ELM
RMSE	4.874×10^{-4}	3.409×10^{-4}	2.432×10^{-4}
Accuracy	60%	80%	90%

to 90% with error <0.02 and it is 80% with error <0.01. Meanwhile, it can be seen that the root mean square error between the output value with ELM-EMA and the real value is 2.432×10^{-4}, which is significantly better than that of hybrid model with single ELM and single ELM "black box" model. The result indicates that the accuracy of the endpoint carbon content prediction model is satisfied for the actual production.

Conclusions

In this paper, a hybrid intelligent modeling method combining intelligent algorithms and the mechanism model is presented to predict the endpoint carbon content of the molten steel in EAF steelmaking. The mechanism model is based on the analysis of the material balance of the process. The ELM is adopted to predict the parameters which are estimated inaccurately by using traditional mechanism methods, and the EMA is employed to promote the global optimization of the model by optimizing the relevant parameters of the ELM. This hybrid intelligent model is used in the 50t EAF and the experiments demonstrate that both the generalization performance and the prediction accuracy of the model are improved by the proposed prediction model.

Acknowledgements The authors would like to express their thanks for the support by the National Nature Science Foundation of China (NO. 51474024 and NO. 51574021) and the National Key Technology R&D Program of the 12th Five-Year Plan (12FYP2015BAF03B01).

References

1. Zhou GY, Peng ZS (2012) Research and application of final carbon control in a 45t electric arc furnace. J Univ Sci Technol Beijing S1: 49–53 (in Chinese)
2. Rajesh N, Khare MR, Pabi SK (2010) Feed forward neural network for prediction of end blow oxygen in LD converter steel making. Mater Res 1:15–19.
3. Tian HX, Mao ZZ, Wang Y (2008) Hybrid modeling of molten steel temperature prediction in LF. ISIJ Int 1:58–62
4. Jiang J, Sun HD, Sun T, Jiang L (2007) Neural network predictive model for end point temperature in oxidation of 100t EAF. Steelmaking 2:31–33 (in Chinese)
5. Huang GB, Zhu QY Siew CK (2007) Extreme learning machine: theory and applications. Neruocomputing 1–3:489–501
6. Huang GB, Wang DH, Lan Y (2011) Extreme learning machines: a survey. Int J Mach Learn Cyber 2:107–122
7. Han M, Liu C (2014) Endpoint prediction model for basic oxygen furnace steel-making based on membrane algorithm evolving extreme learning machine. Appl Soft Comput 1:430–437
8. He CL, Zhu R, Dong K, Qiu YQ, Sun KM (2011) Modeling of an Impinging oxygen jet on molten bath surface in an 150t electric arc furnace. J Iron Steel Res Int 9:13–20
9. Ma G, Zhu R, Dong K, Li Z, Liu R, Yang L, Wei G (2016) Development and application of electric arc furnace combined blowing technology. Ironmaking Steelmaking 8:594–599
10. Wei GS, Zhu R, Dong K, Ma GH, Cheng T (2016) Research and Analysis on the physical and chemical properties of molten bath with bottom-blowing in EAF steelmaking process. Metall Mater Trans B 5:3066–3079.
11. Păun G (2000) Computing with membranes. J Comput Syst Sci 1:108–143
12. Suresh S, Venkatesh Babu R, Kim H (2009) No-reference image quality assessment using modified extreme learning machine classifier. Appl Soft Comput 2:541–552.
13. Huang GB, Chen L (2008) Enhanced random search based incremental extreme learning machine. Neruocomputing 16–18 3460-3468

DEM Simulation of Dispersion of Cohesive Particles by Spontaneous Inter-particle Percolation in a 3D Random Packed Bed

Heng Zhou, Sheng-li Wu, Ming-yin Kou, Shun Yao, Bing-jie Wen, Kai Gu and Feng Chang

Abstract Spontaneous inter-particle percolation is a common phenomenon in nature and industries. Dispersion of cohesionless particles has been investigated by means of various physical and numerical experiments. However, many granular materials are in cohesive or wet state in metallurgical and mineral processes, and the previous works have a lack of understanding the effect of cohesive force on particle dispersion behaviour. Thus, the present work systematically studies the dispersion of cohesive particles by spontaneous inter-particle percolation in a packed bed using discrete element method (DEM). The results indicate that the vertical velocity of percolating particles increases with increasing the cohesive force from 0 to 2 mg (the gravity force of percolating particle, given by $\rho g \pi d^3/6$). While for a higher cohesive force, e.g. $f_c = 8$ mg, insufficient percolation occurs and percolating particles stick in the packed bed. Percolating particles in the packed bed shows a diffusivity for the cases of smaller cohesive force, and the diffusion property can be described by Einstein-Smoluchowski equation. The transverse dispersion of $f_c =$ 2 mg is smaller than that of $f_c = 0$, while the longitudinal dispersion becomes larger when the cohesive force changes from 0 to 2 mg. This study provides a fundamental understanding on dispersion behaviour of cohesive particles in a packed bed, and is useful for processes understanding and optimization in cohesive particles handling and mixing.

Keywords Inter-particle percolation · Dispersion coefficient · Cohesive force DEM

H. Zhou (✉) · S. Wu · M. Kou · S. Yao · B. Wen · K. Gu · F. Chang
School of Metallurgical and Ecological Engineering, University of Science and Technology Beijing, Beijing 100083, China
e-mail: zhouheng@ustb.edu.cn

G. Lambotte et al. (eds.), *Materials Processing Fundamentals 2018*,
The Minerals, Metals & Materials Series,
https://doi.org/10.1007/978-3-319-72131-6_20

Introduction

Granular materials are commonly present in various processes in chemical engineering, metallurgical, and mineral industries. If the grain sizes in a granular material are very different, one may observe that the smaller grains can drain through the piling of larger ones. This is usually termed spontaneous inter-particle percolation. The percolation is a common phenomenon in nature and industries [1]. One specific industry example can be found in ironmaking blast furnace and COREX shaft furnace. When the smaller iron ore particles are loaded upon larger particles, the smaller particles may pass through the larger ones in descending motion under gravity [2, 3]. In many industrial processes, granular materials may in cohesive or wet state [4–6], and cohesion is the attractive force between particles. If cohesion is significant, substantial difference from the free-flowing behaviour of particulate systems is evident. Therefore, the dispersion of cohesive particles due to percolation process is different from that of cohesionless particles. Thus, it is necessary to study the cohesive particle dispersion phenomenon in a packed bed.

In the past, the phenomena of cohesionless particle percolation have been investigated by means of various physical and numerical experiments. Bridgwater et al. [7–9] made a pioneering research in experimentally studying the inter-particle percolation including percolation velocity, residence time and radial distance distribution. Ippolito et al. [10] experimentally investigated the dispersion of small spherical beads moving under the effect of gravity inside a packing of large spheres. Richard et al. [11] used Monte Carlo method to analyse various properties relevant to percolation. Lomine and Oger [12, 13] conducted experiments and discrete element method (DEM) studies to analyse dispersion of particles through a porous structure. Rahman and Zhu [14, 15] and Li et al. [16] studied the effect of particle properties on particle percolation behaviour in a packed bed using DEM. While these studies provided useful information at microscopic level, some important phenomena have not been understood, for example, the effect of cohesion between particles on percolation behaviour was not studied in details. This is especially important in ironmaking processes. For example, the significant stickiness between iron ore particles was observed within the temperature range of 600–675 °C [17]. In the lump zone of blast furnace or COREX shaft furnace, the cohesion between particles, induced by reduced elastic modulus of relevant materials and in particular, softening-sticking bridge as a result of high temperature operation and reduction reaction of iron ore, may directly affect the dispersion behaviour in the packed bed. But this was not considered in the past studies.

In this work, we will present results of our DEM study of the dispersion of cohesive particles by spontaneous inter-particle percolation in a 3D random packed bed. The effect of cohesive force on percolation velocity, residence time distribution, longitudinal and transverse dispersion is studied. All the findings are useful for processes understanding and optimization in cohesive particles handling and mixing.

Model Details

DEM Model

Each single particle in a considered system undergoes both translational and rotational motion, described by Newton's 2nd law of motion. The forces and torques considered include those originating from the particle's contact with neighbouring particles, walls and surrounding fluids. The governing equations for translation and rotational motions of particle i with R_i, mass m_i and moment of inertia $\boldsymbol{I}_i$ can be written as

$$m_i \mathrm{d}\boldsymbol{v}_i/\mathrm{d}t = \sum_{j=1}^{k_i} (\boldsymbol{f}_{\mathrm{c},ij} + \boldsymbol{f}_{\mathrm{d},ij} + \boldsymbol{f}_{\mathrm{e},ij}) + m_i \boldsymbol{g} \tag{1}$$

$$\boldsymbol{I}_i \mathrm{d}\omega_i/\mathrm{d}t = \sum_{j=1}^{k_i} (\boldsymbol{T}_{ij} + \boldsymbol{M}_{ij}) \tag{2}$$

All the forces and torques used in the model are listed in Table 1 and the parameters used in equations are shown in Table 2. Cohesive force between particles can originate from several sources including van der Waals force, electrostatic force, and liquid bridges (capillary forces). This work is focused on a general understanding of dispersion of cohesive particle in a packed bed, thus a simplified model is employed to reduce computational requirement while reasonable and general results can be obtained. The cohesive force $\boldsymbol{f}_{\mathrm{c},ij}$, whose magnitude is set to be proportional to the gravity force of a fine particle (mg, given by $\rho g \pi d^3/6$), is assumed when the gap between two particles is less than a critical value. The similar approach has been used in previous works for simplification [18, 19].

Simulation Conditions

As shown in Fig. 1, the simulation setup is made of a cylindrical container of $\varphi 15D \times 15D$ filled with a packing of monosize large particles of diameter D (which are referred to as packing particles here). This packing is built by random gravity deposition of particles. This procedure gives a reproducible porosity around 0.4. Small particles (percolating particles) of diameter d are put on the top of the packed bed. They are generated randomly at the centreline of the column in a circle of diameter of $1D$. These percolating particles pass through the packed bed towards the bottom of the column under gravity. Their dynamic details are recorded for analysis. The parameters used in the present simulations are listed in Table 3. In this work, the same simulation process is repeated three times and each packing is rebuilt for each simulation. Each numerical data which are presented in this paper are coming from a statistical mean of several simulations.

Table 1 Components of forces and torques acting on particle i

Force and torque		Symbol	Equation
Normal	Contact force	$\boldsymbol{F}_{\text{cn},ij}$	$-4/3E^{*}\sqrt{R^{*}}\delta_{n}^{3/2}\hat{\boldsymbol{n}}$
	Damping force	$\boldsymbol{F}_{\text{dn},ij}$	$-\eta_{\text{n}}(6m_{ij}E^{*}\sqrt{R^{*}\delta_{n}})^{1/2}\boldsymbol{v}_{\mathbf{n},ij}$
Tangential	Contact force	$\boldsymbol{F}_{\text{ct},ij}$	$-\mu_{\text{s}}\lvert\boldsymbol{F}_{\text{cn},ij}\rvert(1-(1-\delta_{\text{t},ij}/\delta_{\text{t},ij,\max})^{3/2})\hat{\boldsymbol{\delta}}_{\text{t}} \quad (\delta_{\text{t},ij}<\delta_{\text{t},ij,\max})$
	Damping force	$\boldsymbol{F}_{\text{dt},ij}$	$-\eta_{s}(6\mu_{s}m_{ij}\lvert\boldsymbol{F}_{cn,ij}\rvert\sqrt{1-\delta_{t,ij}/\delta_{t,ij,\max}}/\delta_{t,ij,\max})^{1/2}\boldsymbol{v}_{t,ij} \quad (\delta_{t,ij}<\delta_{t,ij,\max})$
Friction force		$\boldsymbol{F}_{\text{t},ij}$	$-\mu_{\text{s}}\lvert\boldsymbol{F}_{\text{cn},ij}\rvert\hat{\boldsymbol{\delta}}_{\text{t}} \quad (\delta_{\text{t},ij}>\delta_{\text{t},ij,\max})$
Gravity		$\boldsymbol{F}_{\text{g},i}$	$m_{i}\boldsymbol{g}$
Tangential torque		$\boldsymbol{T}_{ij}$	$\boldsymbol{R}_{ij}\times(\boldsymbol{F}_{\text{ct},ij}+\boldsymbol{F}_{\text{dt},ij})$
Rolling friction torque		$\boldsymbol{M}_{ij}$	$\mu_{\text{r},ij}\lvert\boldsymbol{F}_{\text{cn},ij}\rvert\hat{\boldsymbol{\omega}}_{\text{t},ij}^{n}$

Note $\frac{1}{R^{*}}=\frac{1}{\lvert\boldsymbol{R}_i\rvert}+\frac{1}{\lvert\boldsymbol{R}_j\rvert}$, $E^{*}=\frac{E}{2(1-\nu^{2})}$, $\hat{\boldsymbol{n}}=\frac{\boldsymbol{R}_i}{\lvert\boldsymbol{R}_i\rvert}$, $\hat{\boldsymbol{\omega}}_{\text{t},ij}=\frac{\boldsymbol{\omega}_{\text{t},ij}}{\lvert\boldsymbol{\omega}_{\text{t},ij}\rvert}$, $\hat{\boldsymbol{\delta}}_{\text{t}}=\frac{\boldsymbol{\delta}_{\text{t}}}{\lvert\boldsymbol{\delta}_{\text{t}}\rvert}$, $\delta_{t,ij,\max}=\mu_{s}\frac{2-\nu}{2(1-\nu)}\delta_{n}$, $\boldsymbol{v}_{ij}=\boldsymbol{v}_j-\boldsymbol{v}_i+\boldsymbol{\omega}_j\times\boldsymbol{R}_j-\boldsymbol{\omega}_i\times\boldsymbol{R}_i$, $\boldsymbol{v}_{n,ij}=(\boldsymbol{v}_{ij}\cdot\hat{\boldsymbol{n}})\cdot\hat{\boldsymbol{n}}$, $\boldsymbol{v}_{t,ij}=(\boldsymbol{v}_{ij}\times\hat{\boldsymbol{n}})\times\hat{\boldsymbol{n}}$

Table 2 Parameters used in equations

Parameters		*Parameters*	
E	Young's modulus	$\hat{\boldsymbol{n}}$	Unit vector from the centre of particle to the contact point
E^*	Equivalent Young's modulus	$\boldsymbol{R}_{ij}$	Vector from the mass centre of particle i to particle j
$f_{c,ij}$	Cohesive force	$\boldsymbol{v}_{t,ij}$	Tangential relative velocity of particle i and j
$f_{d,ij}$	Damp force	ω_i	Rotational velocity
$f_{e,ij}$	Elastic force	R^*	Equivalent radius
g	Gravitational acceleration	μ_r	Rolling friction coefficient
$\boldsymbol{I}_i$	Rotational inertia	δ_n	Normal particle overlap
$\boldsymbol{M}_{ij}$	Rolling friction torque	δ_t	Tangential overlap
m_{ij}	Equivalent mass	$\delta_{t,ij,max}$	Maximum particle tangential overlap
$\hat{\omega}_{t,ij}$	Unit vector of particle angular velocity	$\hat{\boldsymbol{\delta}}_t$	Unit vector of particle tangential overlap

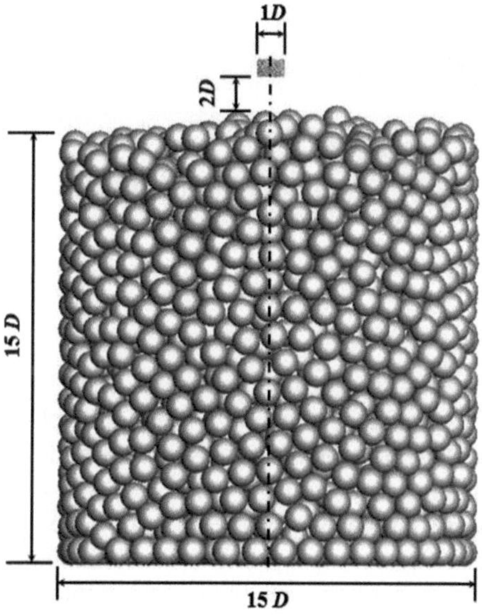

Fig. 1 Geometry of the model used in this work

Results and Discussion

Model Validity

In this work, the present DEM model is validated by the angle of repose for coarse spheres. The simulations were carried out under conditions similar to those used in

Table 3 Particle properties and simulation conditions

Variables	Value
Diameter of packed particle, *D*(m)	0.01
Percolating particle diameter, *d*	1/8*D*
Percolating particle number, *N*	500
Sliding frictional coefficient, μ_s	0.3
Rolling frictional coefficient, μ_r	0.001*D*
Young's modulus, *E*(Pa)	50000 $g\pi D\rho/6$
Poisson's ratio, v_p	0.3
Damping coefficient, *c*	0.3
Time step, Δt(s)	1.0×10^{-7}

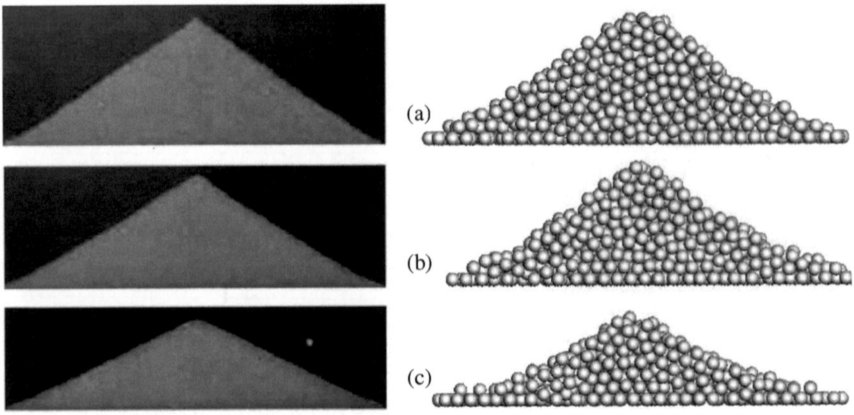

Fig. 2 Sandpiles generated via physical and numerical experiments with different container thickness: **a** $w = 4D$; **b** $w = 12D$; **c** $w = 20D$

previous physical experiments [20]. The so-called discharging method is used to examine the angle of repose. The physical experiments were carried out in a rectangular container with a fixed middle plate and two side outlets. Particles with diameter of 2 mm (particle density = 2500 kg/m^3, $\mu_r = 0.05$ mm and $\mu_s = 0.4$) were employed. Figure 2 shows the typical sandpiles constructed by physical experiments and numerical simulations with different container thicknesses. Because of the relatively small number of particles, the numerical simulation sometimes does not produce a sandpile of smooth straight surface. Nevertheless, the results clearly indicated that the angle of repose decreases with increasing container thickness, and the numerical simulations and physical experiments are comparable. Therefore, the present model is suitable to carry out further simulation.

Effect of Cohesive Force on Percolation Velocity

When a smaller percolating particle is put on a packed bed of larger particles, it may move down through the bed in longitudinal and transverse directions under gravity and interactions with the packing particles. The longitudinal direction is referred to the flow direction, and the transverse one is referred to the direction perpendicular to the flow direction. The percolation velocity reflects, to a degree, the dispersion property of percolating particles. Therefore, the percolation velocity is firstly discussed.

Figure 3 shows the evolution of normalized mean height and normalized mean vertical velocity of percolating particles under different cohesive force conditions. The time is set to be proportional to the free fall time to past a single large sphere diameter and the velocity is in units of free fall velocity reached after falling over one large particle. It is indicated that the normalized height and mean velocity decreases rapidly once percolating particles collide to the packed particles. Then the percolating particles would move toward to the opening of the orifice and percolate among packed particles. For those percolating particles with cohesionless force, the normalized height decreases gradually and the mean velocity decreases progressively towards a steady value. This phenomenon can be related to previous results: for the case of particles falling down in a random packed bed of larger particles, the vertical velocity is a constant [7, 8, 12, 14]. For the cohesive force f_c = 2 mg, the variation trends of the normalized height and mean velocity are similar to that of cohesionless case, except the mean vertical velocity larger than that of f_c = 0 mg. The reason will be discussed in next section by combining the dispersion behaviour. While for the case with f_c = 8 mg, the normalized height decreases progressively towards to a steady value and the mean velocity is reduced to zero. This is because cohesive force is strong enough to resist the inertia motion of percolating particles and the insufficient percolation can occur. Percolating particles will finally adhere to the packed particles and blockage can be observed in this condition.

Figure 4 presents the statistic distributions of residence time for different cohesive forces. As the percolating particles directly stick on the surface of packed particles when the cohesive force f_c = 8 mg, the residence time for this condition is

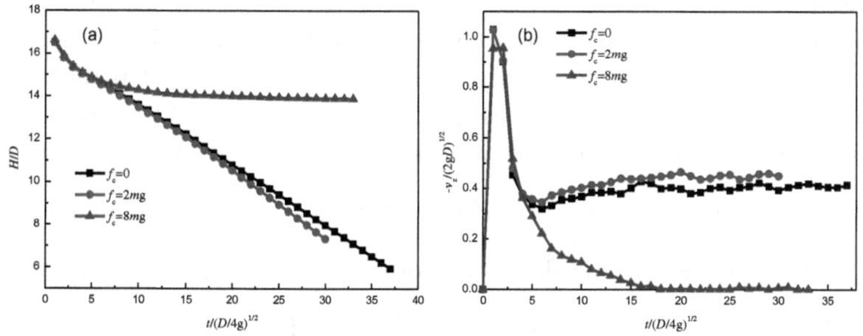

Fig. 3 Evolution of **a** height and **b** velocity of percolating particles for different cohesive forces

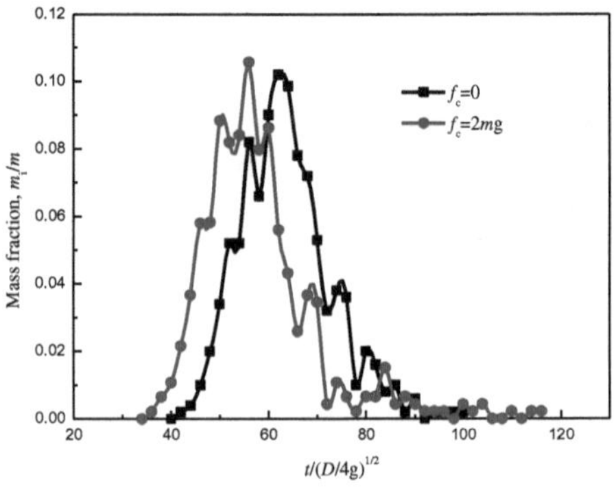

Fig. 4 Statistic distributions of residence time with different cohesive forces

not considered. It has been obtained in the previous experimental studies that the residence time distribution is roughly similar to a normal distribution [8]. Similar trend can also be observed in the present simulation. Besides, with the cohesive force increasing from 0 to 2 mg, the distribution curve shifts to the left. It is because the larger the cohesive force, the higher the particle percolation velocity, and then the less time for the particles to reach the bottom of the packed bed.

Effect of Cohesive Force on Dispersion Behaviour

The dispersion of percolating particles is a random walk process. The percolation flow of the small particle is subject to stochastic motion, which is caused by the interaction between percolating and packing particles. Such stochastic motion leads to dispersion of percolating particle within a region in the packed bed. Figure 5 illustrates the distribution function of particle positions at the exit of packed bed for different cohesive forces. It can be seen that the smaller the cohesive force, the more off-centre the dispersion distance. Especially for the case f_c = 8 mg, the insufficient percolation occurs and the percolating particles almost stick in the centre part of the packed bed. In order to quantitatively describe the dispersion behaviour, detailed information, such as the transverse and longitudinal dispersion coefficients, will be discussed.

The DEM model is possible to access individual particle positions, anywhere at any time, inside the packing of larger spheres. The position of particle k in the horizontal plane can be described as $r_k^2 = x_k^2 + y_k^2$, where x_k and y_k are the particle positions. So, the variance of the position distributions of the N moving particles in this plane is $\left\langle (\Delta r)^2 \right\rangle = \frac{1}{N}\sum_{k=1}^{N} (r_k - \langle r \rangle)^2$, where $\langle r \rangle = \frac{1}{N}\sum_{k=1}^{N} r_k$. In the same manner, if the particle position in the flow direction is denoted by z_k, it can be written as $\left\langle (\Delta z)^2 \right\rangle = \frac{1}{N}\sum_{k=1}^{N} (z_k - \langle z \rangle)^2$, where, $\langle z \rangle = \frac{1}{N}\sum_{k=1}^{N} z_k$. Then, the transverse and axial

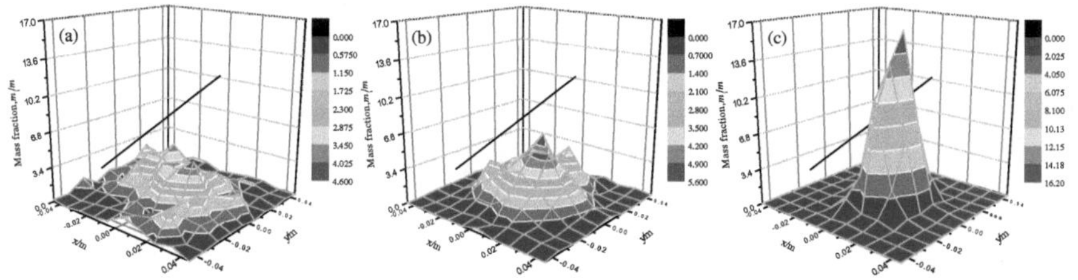

Fig. 5 Distribution function of particle positions at the exit of packed bed for cohesive force **a** $f_c = 0$, **b** $f_c = 2$ mg, **c** $f_c = 8$ mg

dispersion coefficients $D_\perp$ and $D_\parallel$ can be defined from the time evolution of $\left\langle (\Delta r)^2 \right\rangle$ and $\left\langle (\Delta z)^2 \right\rangle$ with Einstein-Smoluchowski equation as follows.

$$\left\langle (\Delta r)^2 \right\rangle = 2D_\perp t \text{ and } \left\langle (\Delta z)^2 \right\rangle = 2D_\parallel t \quad (3)$$

Figure 6 presents the variances, of particle position distribution for different cohesive forces versus time. Linear fits obtained with unweighted least-squares method are presented for cohesive force $f_c = 0$ and $f_c = 2$ mg. We can notice that the linear evolution of the two variances with time is a typical proof of a diffusive property. The deviation from the linear regression observed for smaller values of t is mainly due to the transition phase which occurs in the first few layers of packed bed and the percolating particles need time to reach a diffusive behaviour. On the other hand, for the larger values of t, the deviation from the fits in Fig. 6 can be explained by the finite size of our simulation and some percolating particles have already reached the bottom of the packed bed. Diffusive motion of an isolated particle and a blob of particles were found by Ippolito et al. [10] and Lomine et al. [12], respectively. Our simulations prove the same behaviour for smaller cohesive force. However, as mentioned above, the insufficient percolation occurs for the case $f_c = 8$ mg, and no diffusive motion can be observed. From the Fig. 6, it also can be seen that the transverse dispersion of $f_c = 2$ mg is smaller than that of $f_c = 0$, while the longitudinal dispersion becomes larger when cohesive force changes from 0 to 2 mg. The main reason can be explained as: When percolating particles meet packing particles, they would move downwards and experience multiple collisions. All the contact between the percolating and packing particles would directly decrease the horizontal velocity due to the cohesive force acting as an attractive force. Under the effect of cohesive force, percolating particles would explore laterally the packed bed more difficultly and have a greater probability of passing through the vertical pore. Therefore, the transverse dispersion decreases while the longitudinal dispersion and normalized mean vertical velocity increase with the cohesive force changing from 0 to 2 mg.

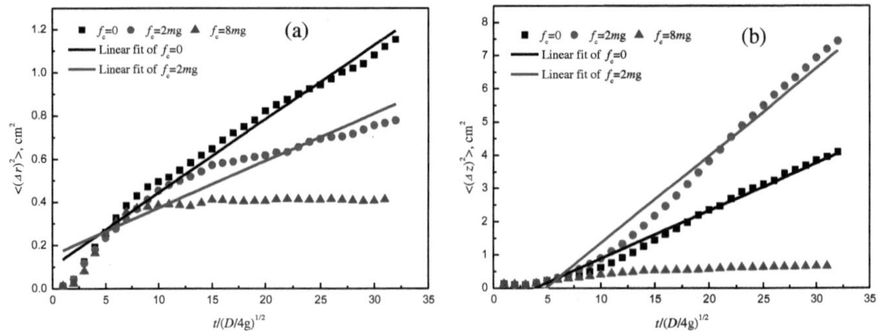

Fig. 6 Time evolutions of **a** $\left\langle(\Delta r)^2\right\rangle$, and **b** $\left\langle(\Delta z)^2\right\rangle$ for different cohesive forces

Conclusions

The percolation behaviour, such as percolation velocity, residence time distribution, longitudinal and transverse dispersion, of cohesive fine particles in a packed bed is studied numerically by means of DEM approach. The conclusion is obtained: The vertical velocity of percolating particles moving down through a random packed bed of larger particles is constant, but it increases with increasing the cohesive force from 0 to 2 mg. While for a higher cohesive force, e.g. f_c = 8 mg, insufficient percolation occurs and percolating particles may stick in the packed bed. The percolating particles moving in the packed bed shows a diffusive property for smaller cohesive force. The transverse dispersion of f_c = 2 mg is smaller than that of f_c = 0, while the longitudinal dispersion becomes larger when cohesive force changes from 0 to 2 mg.

Acknowledgements The authors would like to thank China Postdoctoral Science Foundation (Grant No. 2017M610769) and Fundamental Research Funds for the Central Universities (Grant No. FRF-TP-17-036A1) for their financial supports.

References

1. Rosato A, Strandburg KJ, Prinz F, Swendsen RH (1987) Why the Brazil nuts are on top: size segregation of particulate matter by shaking. Phys Rev Lett 58:1038–1040
2. Yu YW, Westerlund A, Paananen T, Saxén H (2011) Inter-particle percolation segregation during burden descent in the blast furnace. ISIJ Int 51:1050–1056
3. Zhou H, Luo ZG, Zhang T, You Y, Zou ZS, Shen YS (2016) DEM study of solid flow in COREX shaft furnace with areal gas distribution beams. ISIJ Int 56:245–254
4. Samadani A, Kudrolli A (2001) Angle of repose and segregation in cohesive granular matter. Phys Rev E 64:051301
5. Tegzes P, Vicsek T, Schiffer P (2002) Avalanche dynamics in wet granular materials. Phys Rev Lett 89:094301

6. Hou QF, Dong KJ, Yu AB (2014) DEM study of the flow of cohesive particles in a screw feeder. Powder Technol 256:529–539
7. Bridgwater J, Sharpe NW, Stocker DC (1969) Particle mixing by percolation. Trans Inst Chem Eng 47:144–199
8. Bridgwater J, Ingram ND (1971) Rate of spontaneous inter-particle percolation. Trans Inst Chem Eng 49:163–169
9. Scott AM, Bridgwater J (1976) Self-diffusion of spherical particles in a simple shear apparatus. Powder Technol 14:177–183
10. Ippolito I, Samson L, Bourles S, Hulin JP (2000) Diffusion of a single particle in a 3D random packing of spheres. Eur Phys J E 3:227–236
11. Richard P, Oger L, Lemaître J, Samson L, Medvedev NN (1999) Application of the Voronoï tessellation to study transport and segregation of grains inside 2D and 3D packings of spheres. Granular Matter 1:203–211
12. Lomine F, Oger L (2006) Transport of small particles through a 3D packing of spheres: experimental and numerical approaches. J Stat Mech Theory Exp P07019
13. Lomine F, Oger L (2009) Dispersion of particles by spontaneous interparticle percolation through unconsolidated porous media. Phys Rev E 79:051307
14. Rahman M, Zhu HP, Yu AB, Bridgwater J (2008) DEM simulation of particle percolation in a packed bed. Particuology 6:475–482
15. Zhu HP, Rahman M, Yu AB, Bridgwater J, Zulli P (2009) Effect of particle properties on particle percolation behaviour in a packed bed. Miner Eng 22:961–969
16. Li J, Yu AB, Bridgwater J, Rough SL (2010) Spontaneous inter-particle percolation: a kinematic simulation study. Powder Technol 203:397–403
17. Zhang T, Lei C, Zhu QS (2014) Reduction of fine iron ore via a two-step fluidized bed direct reduction process. Powder Technol 254:1–11
18. Pandit JK, Wang XS, Rhodes MJ (2006) On Geldart group a behaviour in fluidized beds with and without cohesive interparticle forces: a DEM study. Powder Technol 164:130–138
19. Hou QF, Samman M, Li J, Yu AB (2014) Modeling the gas-solid flow in the reduction shaft of COREX. ISIJ Int 54:1772–1780
20. Zhou YC, Xu BH, Yu AB, Zulli P (2002) An experimental and numerical study of the angle of repose of coarse spheres. Powder Technol 125:45–54

Author Index

G. Lambotte et al. (eds.), *Materials Processing Fundamentals 2018*,
The Minerals, Metals & Materials Series,
https://doi.org/10.1007/978-3-319-72131-6

Subject Index

G. Lambotte et al. (eds.), *Materials Processing Fundamentals 2018*,
The Minerals, Metals & Materials Series,
https://doi.org/10.1007/978-3-319-72131-6

MIX
Papier aus verantwortungsvollen Quellen
Paper from responsible sources
FSC® C105338

If you have any concerns about our products,
you can contact us on
ProductSafety@springernature.com

In case Publisher is established outside the EU,
the EU authorized representative is:
Springer Nature Customer Service Center GmbH
Europaplatz 3, 69115 Heidelberg, Germany

Printed by Libri Plureos GmbH
in Hamburg, Germany